The Best of Technology Writing 2006

DIGITALCULTUREBOOKS is a collaborative imprint of
the University of Michigan Press and the University of Michigan Library
dedicated to publishing innovative work about the social,
cultural, and political impact of new media.

Brendan I. Koerner, Editor

The Best of Technology Writing 2006

THE UNIVERSITY OF MICHIGAN PRESS AND
THE UNIVERSITY OF MICHIGAN LIBRARY
Ann Arbor

Published in the United States of America by
The University of Michigan Press and
The University of Michigan Library
Manufactured in the United States of America
♾ Printed on acid-free paper

2009 2008 2007 2006 4 3 2 1

A CIP catalog record for this book is available from the British Library.

Library of Congress Cataloging-in-Publication Data

The best of technology writing 2006 / edited by Brendan I. Koerner.
p. cm.
ISBN-13: 978-0-472-03195-5 (pbk. : acid-free paper)
ISBN-10: 0-472-03195-3 (pbk. : acid-free paper)
1. Technical writing. I. Koerner, Brendan I.

T11.B463
600—dc22 2006020166

Contents

Brendan I. Koerner
Introduction *1*

Joshua Davis
La Vida Robot *9*
/Wired/

Alex Ross
The Record Effect *28*
/New Yorker/

David Bernstein
When the Sous-Chef Is an Inkjet *45*
/New York Times/

Edward Tenner
The Rise of the Plagiosphere *50*
/Technology Review/

Clive Thompson
The Xbox Auteurs *54*
/New York Times Magazine/

Farhad Manjoo
Throwing Google at the Book *70*
/Salon/

Steven Johnson
Why the Web Is Like a Rain Forest *85*
/Discover/

Lisa Margonelli
China's Next Cultural Revolution *90*
/Wired/

Mike Daisey
The Coil and I *101*
/Slate/

Daniel H. Pink
The Book Stops Here *106*
/Wired/

Joseph Turow
Have They Got a Deal for You *122*
/Washington Post/

Steven Levy
The Trend Spotter *128*
/Wired/

Adam L. Penenberg
The Right Price for Digital Music *142*
/Slate/

Dan Ferber
Will Artificial Muscle Make You Stronger? *146*
/Popular Science/

Richard Waters
Plugged Into It All *156*
/Financial Times/

Jim Rossignol
Sex, Fame, and PC Baangs *169*
/PC Gamer UK/

Daniel Engber
Crying, While Eating 187
/Slate/

Jay Dixit
Cats with 10 Lives 193
/Legal Affairs/

David A. Bell
The Bookless Future 199
/New Republic/

David McNeill
Mr. Song and Dance Man 220
/Japan Focus/

Koranteng Ofosu-Amaah
Cultural Sensitivity in Technology 225
/Koranteng's Toli/

Justin Mullins
Ups and Downs of Jetpacks 238
/New Scientist/

Jesse Sunenblick
Into the Great Wide Open 248
/Columbia Journalism Review/

Evan Ratliff
The Zombie Hunters 264
/New Yorker/

About the Contributors 279

Acknowledgments 285

Brendan I. Koerner

Introduction

Back in the frothiest dot-com days, a magazine dispatched me to write about a wunderkind software tycoon and his burgeoning company. After an adolescence largely spent coding in his bedroom, the young man had parlayed one of his freeware programs into a fortune roughly the size of Tonga's gross domestic product. My task was to figure out what made this fresh-faced genius tick.

It was an excruciatingly boring assignment. I was allowed just 30 minutes with the mastermind himself, who, though perfectly polite, wasn't exactly a scintillating interview. (He spoke with the languid inflection of someone who'd just ingested two spoonfuls of NyQuil.) The rest of the week was taken up with meetings with marketing executives, who crowed about the company's products being as revolutionary as the bread slicer. I struggled to stay awake as they repeatedly tossed around the phrase "return on investment."

The only memorable event came during a tour of the bowels of the company's Stalinist-style headquarters, where the brainiest employees spent endless hours coding new products. I peeked into one cavernous room where a whey-faced kid sat transfixed at his screen. The company spokeswoman who was escorting me around explained that he was the company's resident designer of "compilers," the esoteric programs that translate source code into machine language.

Correctly sensing that this description of Mr. Whey Face's job wasn't suitably poetic, she paused and tried again: "His job is to think like a machine."

Now *that* sounded interesting—a human being who, day in, day out, was paid to impersonate the mentality of a robot. I flirted with the idea of making this humble compiler designer the subject of the article, rather than his bland boss. The prospect of delving into the mind of a person who was blessed with such an odd and vital talent—a Doctor Doolittle of the computer age, as it were—was strangely alluring.

In the end, alas, I punked out and wrote the standard paean to corporate greatness that my editor demanded. But when it came time to pull together the present collection of technology writing, I vowed to keep an eye peeled for stories that recalled that think-like-a-machine piece that never was—stories that may ostensibly be about bits or motherboards but never lose sight of the human element at their core.

Finding tales that satisfy these criteria was a challenge. The vast majority of technology writing is dominated by product specifications and breathless comparisons of one MP3 player to another. Hyperbole is the norm, as writers—often egged along by headline-conscious editors—are prone to declaring the slightest hint of progress as a development that will "change everything!" And then there are the long-winded, impenetrable pieces that may contain a worthwhile nugget somewhere around the 4000th word but ultimately rival Ambien for soporific effects. Circulating amid this flotsam and jetsam of product reviews and Internet policy polemics, however, is some truly outstanding writing. What distinguishes these pieces, whether they be a narrative *Wired* opus or a lighthearted blog posting, is their authors' awareness that technology, for all of its byzantine details, is essentially an expression of human desire. The need to create machines with ever more RAM or processing power is not hardwired into our DNA, but as thousands of years worth of human civilization have proved, we certainly take an instinctual interest in developing tools that can make our lives easier, facilitate communication, and satiate our curiosity about human perfection and its limits. Whether the technology in question is the plow or the $100 laptop, the creative impulse is the same: to banish hardship as best we can.

That said, tech writers must of course avoid the still alarmingly familiar assumptions that all technological progress is innately good and that morality and ethics have no place in the discussion. The finest technology journalists possess the skeptical eye required to cut through the hype and question the long-term social and cultural effects of innovation. Had tech journalism existed a century ago, for example, we might have better foreseen the eventual consequences of the automobile: geopolitical squabbling over oil, environmental degradation, and the evisceration of cities. At the very least, an astute reporter might have lobbied for preserving the trolleys in my native Los Angeles that were ripped out to make way for freeways and that now double as parking lots most rush hours.

Fortunately, as we enter the era of ubiquitous biotechnology, we have a corps of educated writers who can comment on the potential downside of, say, genetically modifying organisms or fiddling with a fetus's DNA to improve its chances of getting into Yale. The best tech writing tackles such thorny issues head-on and in measured tones, never sacrificing accuracy for the sake of a political agenda.

The best tech writing is also frequently read online, rather than in the pages of magazines or newspapers—publications often jokingly referred to as "dead trees." It has been years since I purchased a hard copy of the *New York Times,* for example, yet I still spend an inordinate amount of time combing through its contents—the only difference is that I now do so on a beautifully crisp 15-inch LCD screen, without fear of ink-stained fingertips.

With this shift in reading habits in mind, we decided to take a cue from the open-source movement and let the Web-surfing public participate in this book's nominations process. Our guidelines were loose, to say the least: we asked only that the entrants had been published in 2005, that they not be rote examples of trade journalism, and that they not require an advanced degree in electrical engineering in order to be understood.

Two things quickly impressed us about the work submitted for consideration: the quality of the prose and the diversity of places in which it was published. The grace of the writing was a particular delight, given that tech journalists aren't especially renowned as great wordsmiths. In fact, the tech section of a newspaper or magazine has traditionally been the journalistic equivalent of the outfield on a T-ball team—that is, a place to stash those whose talents fall well short of enviable. But in a world where even technophobic granddads gush about how digital video recorders have changed their lives, there's an increased demand for geek writers who understand subject-verb agreement as well as what a megapixel is.

These writers aren't just working for the usual suspects of tech journalism, such as *Wired, Technology Review,* or the Circuits section of the *New York Times.* They're also spouting off in the pages of *Slate, Salon,* the *New Republic,* and a host of other publications, both dead tree and digital, which are usually associated with political and cultural junkies more than hardcore geeks. And then, of course, there are the bloggers, who churn out some surprisingly well-crafted commentaries in exchange for zero compensation, aside from the warm fuzzies proffered by readers in their comments sections. Though I confess that this doesn't bode well for my future financial prospects, many of these unpaid writers produce more perceptive, well-informed pieces than my colleagues who make a living by opining on the social impact of Blackberries.

Ultimately, though, the majority of pieces selected for this volume came from the more traditional standard-bearers of technology journalism. One favorite is *Wired* contributing editor Joshua Davis's "La Vida Robot," the absolutely enthralling tale of four Mexican American teenagers—each of them an undocumented immigrant—who banded together to build a killer underwater robot. Davis doesn't skimp on the technical details of the machine the quartet constructed, describing in depth, for example, the team's decision to use PVC pipe in lieu of foam. The real joy of the article, however, is the way in which Davis fleshes out each character, so that by the end our hearts sink upon learning that Oscar Vazquez, the team's leader, is hanging Sheetrock rather than attending college. "La Vida Robot" may be a *Rocky*-like yarn, but it's also a painful reminder of our nation's myopia about the importance of educating engineers. If there's any justice in this world, Vazquez will get that degree and end up working on NASA's Mars mission someday.

Several strong pieces were recommended from *Slate,* including Daniel Engber's hilarious account of meme creation, "Crying, While Eating: My Sad, Hungry Climb to Internet Stardom." An entrant in a contest to see who could create the most visited Web site from scratch, Engber and his colleagues came up with a truly off-the-wall idea: a collection of videos showing various folks, um, crying while eating. (Example: Engber himself sobbed while scarfing down *soba,* ostensibly because he "ruined Passover.") What ensues is a whirlwind trip through the online universe, as word of the odd site spreads from BoingBoing to Norwegian-language blogs in record time. It's an insightful lesson on mob psychology, not to mention the effects of cheap, ubiquitous DSL.

The availability of all that low-priced bandwidth is undoubtedly changing the way we process information, and several of our contributors focused on what's next for old-school mediums such as books, newspapers, and even clipped-out recipes. Take David A. Bell's "The Bookless Future," an examination of the post–printing press world by a distinguished scholar of French history, which first ran in the *New Republic.* Peppering his narrative with personal anecdotes about his Napoleonic research, Bell manages to explore a wonkish issue with just the right modicum of wit. He obviously knows his gadgets, too, and his expert deconstruction of what's gone wrong with e-readers so far will please bookworms and gearheads alike.

And that is really the hallmark of a great bit of technology writing—something that you needn't be a bona fide member of the digerati to enjoy and, more important, "get." There were many promising nominees that passed through the transom but were ultimately judged too esoteric for inclusion. With all due respect and love to Slashdot, the Google of geekdom, we didn't want writing for the Slashdot crowd; we wanted writing that might bring a mainstream audience a little closer to (someday) appreciating Slashdot itself.

That meant seeking out nuanced writers who could think like machines when need be but could also express themselves in the vernacular. In practice, that also meant accepting some works from writers who were more than a bit surprised to hear themselves described as tech journalists. Among those who were slightly flabbergasted to be selected for a technology collection was the *New Yorker*'s Alex Ross, whose "The Record Effect" describes how the invention of the phonograph forever altered the way music was composed and performed. At first glance, Ross's piece does seem to have more in common with intellectual history than, say, a *PC Magazine* laptop review. But in between his learned comments on Chuck Berry and Arturo Toscanini, Ross deftly comments on the unstoppable MP3ing of the world's music: "Would Beethoven or Billie [Holiday] ever have existed," he asks, "if people had always listened to music the way we listen now?"

I'll leave it to you to discover Ross's answer, just as I'll leave it to you to discover why *Legal Affairs*' Jay Dixit believes that feline cloning needs to be regulated, why blogger Koranteng Ofosu-Amaah thinks that Yahoo's photo technology stacks the deck against Africans, and how *Japan Focus*'s David McNeill came to meet the humble (and seriously undercompensated) inventor of the karaoke machine. All of these revelations are so entertaining, you'll probably forget that, yes, you're reading technology journalism, a genre that is too rarely associated with enjoyable reads. And that's exactly the point. No one can dispute that tech writing is important, given technology's central role in reshaping the world around us. But being important shouldn't condemn it to dour and lifeless prose. Just as writers like James B. Stewart (*Den of Thieves*) and Michael Lewis (*Liar's Poker*) once proved that even business journalism could display literary flair, so the authors in this book demonstrate that tech writing too can be beguiling, compassionate, and graceful.

Here's hoping that the stories in this volume inspire other skilled writers to try their hand at writing about technology, regardless of whether or not they can tell HTML from XML or a microchip from a nanochip. Though a dash

Joshua Davis

La Vida Robot

How four underdogs from the mean streets of Phoenix took on the best from MIT in the national underwater bot championship

The winter rain makes a mess of West Phoenix. It turns dirt yards into mud and forms reefs of garbage in the streets. Junk food wrappers, diapers, and porn in Spanish are swept into the gutters. On West Roosevelt Avenue, security guards, two squad cars, and a handful of cops watch teenagers file into the local high school. A sign reads: Carl Hayden Community High School: The Pride's Inside.

There certainly isn't a lot of pride on the outside. The school buildings are mostly drab, late '50s–era boxes. The front lawn is nothing but brown scrub and patches of dirt. The class photos beside the principal's office tell the story of the past four decades. In 1965, the students were nearly all white, wearing blazers, ties, and long skirts. Now the school is 92 percent Hispanic. Drooping, baggy jeans and XXXL hoodies are the norm.

The school PA system crackles, and an upbeat female voice fills the bustling linoleum-lined hallways. "Anger management class will begin in five minutes," says the voice from the administration building. "All referrals must report immediately."

Across campus, in a second-floor windowless room, four students huddle around an odd, three-foot-tall frame constructed of PVC pipe. They have equipped it with propellers, cameras, lights, a laser, depth detectors, pumps, an underwater microphone, and an articulated pincer. At the top sits a black, waterproof briefcase containing a nest of hacked processors, minuscule fans, and LEDs. It's a cheap but astoundingly functional underwater robot capable of recording sonar pings and retrieving objects 15 feet below the surface. The four teenagers who built it are all undocumented Mexican immigrants who came to this country through tunnels or hidden in the backseats of cars. They live in sheds and rooms without electricity. But over three days last summer, these kids from the desert proved they were among the smartest young underwater engineers in the country.

It was the end of June. Lorenzo Santillan, 16, sat in the front seat of the school van and looked out at the migrant farmworkers in the fields along Interstate 10. Lorenzo's face still had its baby fat, but he'd recently sprouted a mustache and had taken to wearing a fistful of gold rings, a gold chain, and a gold medallion of the Virgin Mary pierced through the upper part of his left ear. The bling wasn't fooling anyone. His mother had been fired from her job as a hotel maid, and his father had trouble paying the rent as a gardener. They were on the verge of eviction for nonpayment.

He could see himself having to quit school to work in those fields.

"What's a PWM cable?" The sharp question from the van's driver, Allan Cameron, snapped Lorenzo out of his reverie. Cameron is the computer science teacher who cosponsors Carl Hayden's robotics program. At 59, he has a neatly trimmed white beard, unkempt brown hair, and more energy than most men half his age. Together with his fellow science teacher Fredi Lajvardi, Cameron had put up flyers around the school a few months earlier, offering to sponsor anyone interested in competing in the third annual Marine Advanced Technology Remotely Operated Vehicle Competition. Lorenzo was one of the first to show up to the after-school meeting last spring.

Cameron hadn't expected many students to be interested, particularly not a kid like Lorenzo, who was failing most of his classes and perpetually looked like he was about to fall asleep. But Lorenzo didn't have much else to do after school. He didn't want to walk around the streets. He had tried that—he'd been a member of WBP 8th Street, a West Side gang. When his friends started to get arrested for theft, he dropped out. He didn't want to go to jail.

That's why he decided to come to Cameron's meeting.

"PWM," Lorenzo replied automatically from the van's passenger seat. "Pulse width modulation. Esto controls analog circuits with digital output."

Over the past four months, Lorenzo had flourished, learning a new set of acronyms and raising his math grade from an F to an A. He had grown up rebuilding car engines with his brother and cousin. Now he was ready to build something of his own. The team had found its mechanics man.

Ever since his younger sister demanded her own room four years ago, Cristian Arcega has been living in a 60-square-foot plywood shed attached to the side of his parents' trailer. He likes it there. It's his own space.

He's free to contemplate the acceleration of a raindrop as it leaves the clouds above him. He can hear it hit the roof and slide toward the puddles on the street outside. He imagines that the puddles are oceans and that the underwater robot he was building at school can explore them.

Cameron and Ledge, as the students called Lajvardi, formed the robotics group for kids like Cristian. He was probably the smartest 16-year-old in West Phoenix—without even trying, he had one of the highest GPAs in the school district. His brains and diminutive stature (five feet four, 135 pounds) kept him apart at Carl Hayden. That and the fact that students socialized based on Mexican geography: In the cafeteria, there were Guanajuato tables and Sonoran tables. Cristian was from Mexicali, but he'd left Mexico in the back of a station wagon when he was 6. He thought of himself as part American, part Mexican, and he didn't know where to sit.

So he ate lunch in the storage room the teachers had comandeered for the underwater remotely operated vehicle (ROV) club. Cristian devoted himself to solving thrust vector and power supply issues. The robot competition required students to build a bot that could survey a sunken mock-up of a submarine (it was sponsored in part by the Office of Naval Research and NASA)—not easy stuff. The teachers had entered the club in the expert-level Explorer class instead of the beginner Ranger class. They figured their students would lose anyway, and there was more honor in losing to the college kids in the Explorer division than to the high schoolers in Ranger. Their real goal was to show the students that there were opportunities outside West Phoenix. The teachers wanted to give their kids hope.

Just getting them to the Santa Barbara contest in June with a robot would be an accomplishment, Cameron thought. He and Ledge had to gather a group of students who, in four months, could raise money, build a robot, and learn how to pilot it. They had no idea they were about to assemble the perfect team.

"We should use glass syntactic flotation foam," Cristian said excitedly at that first meeting. "It's got a really high compressive strength."

Cameron and Ledge looked at each other. Now they had their genius.

Oscar Vazquez was a born leader. A senior, he'd been in ROTC since ninth grade and was planning on a career in the military. But when he called to schedule a recruitment meeting at the end of his junior year, the officer in charge told him he was ineligible for military service. Because he was an undocumented alien—his parents had brought him to the United States from Mexico when he was 12—he couldn't join, wouldn't get any scholarships, and had to start figuring out what else to do with his life. Oscar felt aimless until he heard about the robot club from Ledge, who was teaching his senior biology seminar. Maybe, he thought, engineering could offer him a future.

ROTC had trained Oscar well: He knew how to motivate people. He made sure that everyone was in the room and focused when he phoned Frank Szwankowski, who sells industrial and scientific thermometers at Omega Engineering in Stamford, Connecticut. Szwankowski knows as much about thermometer applications as anyone in the United States. All day long, he talks to military contractors, industrial engineers, and environmental consultants. So he was momentarily confused when he heard Oscar's high-pitched Mexican accent on the other end of the line. The 17-year-old kid from the desert wanted advice on how to build a military-grade underwater ROV.

This was the second call Szwankowski had received

from amateur roboticists in less than a month. A few weeks earlier, students from MIT's oceanic engineering department had called and said they were entering the national underwater ROV championships. Oscar said that his team, too, was competing and needed to learn as much as it could from the experts. Szwankowski was impressed. The MIT kids had simply asked him what they wanted and hung up.

Oscar spent 45 minutes on the phone digging deeper and deeper into thermometer physics.

Oscar began by explaining that his high school team was taking on college students from around the United States. He introduced his teammates: Cristian, the brainiac; Lorenzo, the *vato loco* who had a surprising aptitude for mechanics; and 18-year-old Luis Aranda, the fourth member of the crew. At five feet ten and 250 pounds, Luis looks like Chief from *One Flew over the Cuckoo's Nest.* He was the tether man, responsible for the pickup and release of what would be a 100-pound robot.

Szwankowski was impressed by Oscar. He launched into an in-depth explanation of the technology, offering details as if he were letting them in on a little secret. "What you really want," he confided, "is a thermocouple with a cold junction compensator." He went over the specifications of the device and then paused. "You know," he said, "I think you can beat those guys from MIT. Because none of them know what I know about thermometers."

"You hear that?" Oscar said triumphantly when they hung up. He looked at each team member pointedly. "We got people believing in us, so now we got to believe in ourselves."

Oscar helped persuade a handful of local businesses to donate money to the team. They raised a total of about $800. Now it was up to Cristian and Lorenzo to figure out what to do with the newfound resources. They began by sending Luis to Home Depot to buy PVC pipe. Despite the donations, they were still on a tight budget. Cristian would have to keep dreaming about glass syntactic flotation foam; PVC pipe was the best they could afford.

But PVC had benefits. The air inside the pipe would create buoyancy as well as provide a waterproof housing for wiring. Cristian calculated the volume of air inside the pipes and realized immediately that they'd need ballast.

He proposed housing the battery system on board, in a heavy waterproof case.

It was a bold idea. If they didn't have to run a power line down to the bot, their tether could be much thinner, making the bot more mobile. Since the competition required that their bot run through a series of seven exploration tasks—from taking depth measurements to locating and retrieving acoustic pingers—mobility was key. Most of the other teams wouldn't even consider putting their power supplies in the water. A leak could take the whole system down. But if they couldn't figure out how to waterproof their case, Cristian argued, then they shouldn't be in an underwater contest.

While other teams machined and welded metal frames, the guys broke out the rubber glue and began assembling the PVC pipe. They did the whole thing in one night, got high on the pungent fumes, and dubbed their new creation Stinky. Lorenzo painted it garish shades of blue, red, and yellow to designate the functionality of specific pipes. Every inch of PVC had a clear purpose. It was the type of machine only an engineer would describe as beautiful.

Carl Hayden Community High School doesn't have a swimming pool, so one weekend in May, after about six weeks of work in the classroom, the team took Stinky to a scuba training pool in downtown Phoenix for its baptism.

Luis hefted the machine up and gently placed it in the

water. They powered it up. Cristian had hacked together off-the-shelf joysticks, a motherboard, motors, and an array of onboard finger-sized video cameras, which now sent flickering images to black-and-white monitors on a folding picnic table.

Using five small electric trolling motors, the robot could spin and tilt in any direction. To move smoothly, two drivers had to coordinate their commands. The first thing they did was smash the robot into a wall.

"This is good, this is good," Oscar kept repeating, buying himself a few seconds to come up with a positive spin. "Did you see how hard it hit the wall? This thing's got power. Once we figure out how to drive it, we'll be the fastest team there."

By early June, as the contest neared, the team had the hang of it. Stinky now buzzed through the water, dodging all obstacles. The drivers, Cristian and Oscar, could make the bot hover, spin in place, and angle up or down. They could send enough power to Stinky's small engines to pull Luis around the pool. They felt like they had a good shot at not placing last.

The team arrived at the Olympic-size UC Santa Barbara pool on a sunny Thursday afternoon. The pool was concealed under a black tarp—the contest organizers didn't want the students to get a peek at the layout of the mission. College students from cities across the country—Long Beach, California; Miami, Florida; Galveston, Texas; New Haven, Connecticut; and half a dozen others—milled around the water's edge. The Carl Hayden teammates tried to hide their nervousness, but they were intimidated. Lorenzo had never seen so many white people in one place. He was also new to the ocean.

He had seen it for the first time several months earlier

on a school trip to San Diego. It still unnerved him to see so much water. He said it was "incredifying"—incredible and terrifying at the same time.

Even though Lorenzo had never heard of MIT, the team from Cambridge scared him, too. There were 12 of them—6 ocean-engineering students, 4 mechanical engineers, and 2 computer science majors. Their robot was small, densely packed, and had a large ExxonMobil sticker emblazoned on the side. The largest corporation in the United States had kicked in $5,000 for the privilege of sponsoring them. Other donations brought the MIT team's total budget to $11,000.

As Luis hoisted Stinky to the edge of the practice side of the pool, Cristian heard repressed snickering. It didn't give him a good feeling. He was proud of his robot, but he could see that it looked like a Geo Metro compared with the Lexuses and BMWs around the pool. He had thought that Lorenzo's paint job was nice. Now it just looked clownish.

Things got worse when Luis lowered Stinky into the water. They noticed that the controls worked only intermittently. When they brought Stinky back onto the pool deck, there were a few drops of water in the waterproof briefcase that housed the control system. The case must have warped on the trip from Arizona in the back of Ledge's truck. If the water had touched any of the controls, the system would have shorted out and simply stopped working.

Cristian knew that they were faced with two serious problems: bad wiring and a leak.

Oscar sketched out the situation. They'd have to resolder every wire going into the main controller in the next 12 hours. And they would either have to fix the leak or find something absorbent to keep moisture away from the onboard circuitry.

An image from television flashed through Lorenzo's mind. "Absorbent?" he asked. "Like a tampon?"

The Ralph's grocery store near the UCSB campus is done up to look like a hacienda, complete with a red tile roof, glaringly white walls, and freshly planted palms. The guys dropped Lorenzo off in front. It was his bright idea, after all. He wandered past the organic produce section, trying to build up his courage. He passed an elderly lady examining eggplant—he was too embarrassed to ask her. Next, he saw a young woman in jeans shopping for shampoo.

"Excuse me, madam," he began. He wasn't used to approaching women, let alone well-dressed white women. He saw apprehension flash across her face.

Maybe she thought he was trying to sell magazines or candy bars, but he steeled himself. He explained that he was building a robot for an underwater contest, and it was leaking. He wanted to soak up the water with tampons but didn't know which ones to buy. "Could you help me buy the most best tampons?"

The woman broke into a big smile and led him to feminine hygiene. She handed him a box of O.B. ultraabsorbency. "These don't have an applicator, so they'll be easier to fit inside your robot," she said. He stared at the ground, mumbled his thanks, and headed quickly for the checkout.

"I hope you win," she called out, laughing.

Someone had to be well rested for the contest, so Cristian and Luis slept that night. Oscar and Lorenzo stayed up resoldering the entire control system. It was nerve-racking work. The wires were slightly thicker than a human hair, and there were 50 of them. If the soldering iron got too close to a wire, it would melt and there'd be no time to rip the

PVC and cable housing apart to fix it. One broken wire would destroy the whole system, forcing them to withdraw from the contest.

By two in the morning, Oscar's eyesight was blurring, but he kept at it.

Lorenzo held the wires in place while Oscar lowered the soldering gun. He dropped one last dab of alloy on the connection and sat back. Lorenzo flipped the power switch. Everything appeared to work again.

On the day of the contest, the organizers purposely made it difficult to see what was happening under the water. A set of high-powered fans blew across the surface of the pool, obscuring the view below and forcing teams to navigate by instrumentation alone. The side effect was that no one had a good sense of how the other teams were doing.

When Luis lowered Stinky into the water for their run, Lorenzo prayed to the Virgin Mary. He prayed that the tampons would work but then wondered if the Virgin got her period and whether it was appropriate for him to be praying to her about tampons. He tried to think of a different saint to pray to but couldn't come up with an appropriate one. The whir of Stinky's propellers brought him back to the task at hand, extracting a water sample from a submerged container.

The task was to withdraw 500 milliliters of fluid from the container 12 feet below the surface. Its only opening was a small, half-inch pipe fitted with a one-way valve. Though the Carl Hayden team didn't know it, MIT had designed an innovative system of bladders and pumps to carry out this task.

MIT's robot was supposed to land on the container, create a seal, and pump out the fluid. On three test runs in Boston, the system worked fast and flawlessly.

MIT's ROV motored smoothly down and quickly located the five-gallon drum inside the plastic submarine mock-up at the bottom of the pool. But as the robot approached the container, its protruding mechanical arm hit a piece of the submarine frame, blocking it from going farther. They tried a different angle but still couldn't reach the drum. The bot wasn't small enough to slip past the gap in the frame, making their pump system useless. There was nothing they could do—they had to move on to the next assignment.

When Stinky entered the water, it careened wildly as it dived toward the bottom. Luis stood at the pool's edge, paying out the tether cable. From the control tent, Cristian, Oscar, and Lorenzo monitored Stinky's descent on their video screens.

"Vamonos, Cristian, this is it!" Oscar said, pushing his control too far forward. They were nervous and overcompensated for each other's joystick movements, causing Stinky to veer off course. When they reached the submarine, they saw the drum and tried to steady the robot. Stinky had a bent copper proboscis, a bilge pump, and a dime-store balloon. They had to fit their long, quarter-inch-wide sampling tube into a half-inch pipe and then fill the balloon for exactly 20 seconds to get 500 milliliters. They had practiced dozen of times at the scuba pool in Phoenix, and it had taken them, on average, 10 minutes to stab the proboscis into the narrow tube.

Now they had 30 minutes total to complete all seven tasks on the checklist.

It was up to Oscar and Cristian. They readjusted their grip on the joysticks and leaned into the monitors. Stinky hovered in front of the submarine framing that had frustrated the MIT team. Because Stinky's copper pipe was 18 inches long, it was able to reach the drum. The control tent was silent. Now that they were focused on the mission, both pilots relaxed and made almost imperceptibly small movements with their joysticks. Oscar tapped the control forward while Cristian gave a short backward blast on the vertical propellers. As Stinky floated forward a half inch, its rear raised up, and the sampling pipe sank perfectly into the drum.

"Dios mio," Oscar whispered, not fully believing what he saw.

He looked at Lorenzo, who had already activated the pump and was counting out 20 seconds in a decidedly unscientific way.

"Uno, dos, tres, quatro . . ."

Oscar backed Stinky out of the sub. They spun the robot around, piloted it back to Luis at the edge of the pool, and looked at the judges, who stood in the control tent behind them.

"Can we make a little noise?" Cristian asked Pat Barrow, a NASA lab operations manager supervising the contest.

"Go on ahead," he replied.

Cristian started yelling, and all three ran out to hug Luis, who held the now-filled blue balloon. Luis stood there with a silly grin on his face while his friends danced around him.

It was a short celebration. They still had four more tasks. Luis attached Szwankowski's thermometer and quickly lowered the ROV back into the water.

Tom Swean is the gruff 58-year-old head of the Navy's Ocean Engineering and Marine Systems program. He develops million-dollar autonomous underwater robots for the SEALs at the Office of Naval Research. He's not used to dealing with Mexican American teenagers sporting gold chains; fake diamond rings; and patchy, adolescent mustaches.

The Carl Hayden team stood nervously in front of him. He stared sullenly at them. This was the engineering review—professionals in underwater engineering evaluated all the ROVs, scored each team's technical documentation, and grilled students about their designs. The results counted for more than half of the total possible points in the contest.

"How'd you make the laser range finder work?" Swean growled. MIT had admitted earlier that a laser would have been the most accurate way to measure distance underwater, but they'd concluded that it would have been difficult to implement.

"We used a helium neon laser, captured its phase shift with a photo sensor, and manually corrected by 30 percent to account for the index of refraction," Cristian answered rapidly, keyed up on adrenaline. Cameron had peppered them with questions on the drive to Santa Barbara, and Cristian was ready.

Swean raised a bushy, graying eyebrow. He asked about motor speed, and Lorenzo sketched out their combination of controllers and spike relays.

Oscar answered the question about signal interference in the tether by describing how they'd experimented with a 15-meter cable before jumping up to one that was 33 meters.

"You're very comfortable with the metric system," Swean observed.

"I grew up in Mexico, sir," Oscar said.

Swean nodded. He eyed their rudimentary flip chart.

"Why don't you have a PowerPoint display?" he asked.

"PowerPoint is a distraction," Cristian replied. "People use it when they don't know what to say."

"And you know what to say?"

"Yes, sir."

In the lobby outside the review room, Cameron and Ledge waited anxiously for the kids. They expected them to come out shaken, but all four were smiling—convinced that they had answered Swean's questions perfectly.

Cameron glanced nervously at Ledge. The kids were too confident. They couldn't have done that well.

Still, both teachers were in a good mood. They had learned that the team placed third out of 11 in the seven underwater exercises. Only MIT and Cape Fear Community College from North Carolina had done better. The overall winner would be determined by combining those results with the engineering interview and a review of each group's technical manual. Even if they did poorly on the interview, they were now positive that they hadn't placed last.

"Congratulations, guys," Cameron said. "You officially don't suck."

"Can we go to Hooters if we win?" Lorenzo asked.

"Sure," Ledge said with a dismissive laugh. "And Dr. Cameron and I will retire, too."

The awards ceremony took place over dinner, and the Carl Hayden team was glad for that. They hadn't eaten well over the past two days, and even flavorless iceberg lettuce looked good to them. Their nerves had calmed.

After the engineering interview, they decided that they had probably placed somewhere in the middle of the pack, maybe fourth or fifth overall.

Privately, each of them was hoping for third.

The first award was a surprise: a judge's special prize that wasn't listed in the program. Bryce Merrill, the bearded, middle-aged recruiting manager for Oceaneering International, an industrial ROV design firm, was the announcer. He explained that the judges created this spontaneously to honor special achievement. He stood behind a podium on the temporary stage and glanced down at his notes. The contestants sat crowded around a dozen tables. Carl Hayden High School, he said, was that special team.

The guys trotted up to the stage, forcing smiles. It seemed obvious that this was a condescending pat on the back, as if to say, "A for effort!"

They didn't want to be "special"—they wanted third. It signaled to them that they'd missed it.

They returned to their seats, and Cameron and Ledge shook their hands.

"Good job, guys," Ledge said, trying to sound pleased. "You did well. They probably gave you that for the tampon."

After a few small prizes were handed out (Terrific Tether Management, Perfect Pick-up Tool), Merrill moved on to the final awards: Design Elegance, Technical Report, and Overall Winner. The MIT students shifted in their seats and stretched their legs. While they had been forced to skip the fluid sampling, they had completed more underwater tasks overall than Carl Hayden or Cape Fear. The Cape Fear team sat across the room, fidgeted with their napkins, and tried not to look nervous. The students from Monterey Peninsula College looked straight ahead. They placed fourth behind Carl Hayden in the underwater trials. They were the most likely third-place finishers. The guys from Phoenix glanced back at the buffet table and wondered if they could get more cake before the ceremony ended.

Then Merrill leaned into the microphone and said that the ROV named Stinky had captured the design award.

"What did he just say?" Lorenzo asked.

"Oh my God!" Ledge shouted. "Stand up!"

Before they could sit down again, Merrill told them that they had won the technical writing award.

"Us illiterate people from the desert?" Lorenzo thought. He looked at Cristian, who had been responsible for a large part of the writing.

Cristian was beaming. To his analytical mind, there was no possibility that his team—a bunch of ESL students—could produce a better written report than kids from one of the country's top engineering schools.

They had just won two of the most important awards. All that was left was the grand prize. Cristian quickly calculated the probability of winning but couldn't believe what he was coming up with. Ledge leaned across the table and grabbed Lorenzo's shirt. "Lorenzo, if what I think is about to happen does happen, I do not, under any circumstances, want to hear you say the word 'Hooters' onstage."

"And the overall winner for the Marine Technology ROV championship," Merrill continued, looking up at the crowd, "goes to Carl Hayden High School of Phoenix, Arizona!"

Lorenzo threw his arms into the air, looked at Ledge, and silently mouthed the word "Hooters."

Cameron and Ledge haven't taken Lorenzo to Hooters, nor have they retired.

They hope to see all four kids go to college before they quit teaching, which means they're likely to keep working for a long time. Since the teenagers are undocumented, they don't qualify for federal loans. And though they've lived in Arizona for an average of 11 years, they would still have to pay out-of-state tuition, which can be as much as three times the in-state cost. They can't afford it.

And they're not alone. Approximately 60,000 undocumented students graduate from U.S. high schools every year. One promising solution, according to Cameron and other advocates for immigrant kids, is the Dream Act, federal legislation that would give in-state tuition and temporary resident status to undocumented students who graduate from a U.S. high school after being enrolled in the States for five years or more. The bill, which was introduced in 2003 and is slated to be resubmitted this spring, aims to give undocumented students a reason to stay in school. If they do, the act promises financial assistance for college. In turn, immigrants would pay taxes and be able to contribute their talents to the United States.

Some immigration activists don't see it that way. Ira Mehlman, the Los Angeles–based media director for the Federation for American Immigration Reform, successfully lobbied against the legislation last year. He says it will put citizens and legal immigrants in direct competition for the limited number of seats at state colleges. "What will you say to an American kid who does not get into a state university and whose family cannot afford a private college because that seat and that subsidy have been given to someone who is in the country illegally?" he asks.

Oscar wipes the white gypsum dust from his face. It's a hot Tuesday afternoon in Phoenix, and he's putting up Sheetrock. He graduated from Carl Hayden last spring, and this is the best work he can find. He enjoys walking into the half-built homes and analyzing the engineering. He thinks it'll keep him sharp until he can save up enough money to study engineering at Arizona State University. It will cost him approximately $50,000 as an out-of-state student. That's a lot of Sheetrocking.

Luis also graduated and is filing papers in a Phoenix Social Security Services office. Cristian and Lorenzo are now juniors. Their families can barely support themselves, let alone raise the money to send their kids to college. Last summer, Cristian's hopes flagged even further when his family was forced to spend $3,000 to replace the decrepit air-conditioning unit in their aluminum trailer. Without AC, the trailer turns into a double-wide oven in the desert heat. His family was that much further from being able to save money to send their child to college.

When Oscar gets home from work that night, he watches the gypsum dust swirl down the sink drain when he washes his hands. He wonders what formulas define a vortex. On the other side of the neighborhood, Cristian lies on his bed and tries to picture the moisture in the clouds above. Rain isn't predicted anytime soon.

Alex Ross

The Record Effect

How technology has transformed the sound of music

Ninety-nine years ago, John Philip Sousa predicted that recordings would lead to the demise of music. The phonograph, he warned, would erode the finer instincts of the ear, end amateur playing and singing, and put professional musicians out of work. "The time is coming when no one will be ready to submit himself to the ennobling discipline of learning music," he wrote. "Everyone will have their ready made or ready pirated music in their cupboards." Something is irretrievably lost when we are no longer in the presence of bodies making music, Sousa said. "The nightingale's song is delightful because the nightingale herself gives it forth."

Before you dismiss Sousa as a nutty old codger, you might ponder how much has changed in the past 100 years. Music has achieved onrushing omnipresence in our world: millions of hours of its history are available on disk; rivers of digital melody flow on the Internet; MP3 players with 10,000 songs can be tucked in a back pocket or a purse. Yet, for most of us, music is no longer something we do ourselves or even watch other people doing in front of us. It has become a radically virtual medium, an art without a face. In the future, Sousa's ghost might say, reproduction will replace production entirely. Zombified listeners will shuffle through the archives of the past, and new music will consist of rearrangements of the old.

Ever since Edison introduced the wax cylinder, in 1877, people have been trying to figure out what recording has done for and to the art of music. Inevitably, the conversation has veered toward rhetorical extremes. Sousa was a pioneering spokesman for the party of doom, which was later filled out by various post-Marxist theorists. In the opposite corner are the technological utopians, who will tell you that recording has not imprisoned music but liberated it, bringing the art of the elite to the masses and the art of the margins to the center. Before Edison came along, the utopians say, Beethoven's symphonies could be heard only in select concert halls. Now CDs carry the man from Bonn to the corners of the earth, summoning forth the million souls he hoped to embrace in his "Ode to Joy." Conversely, recordings gave the likes of Louis Armstrong, Chuck Berry, and James Brown the chance to occupy a global platform that Sousa's idyllic old America, racist to the core, would have denied them. The fact that their records played a crucial role in the advancement of African American civil rights puts in proper perspective the aesthetic debate about whether or not technology has been "good" for music.

I discovered much of my favorite music through LPs and CDs, and I am not about to join the party of Luddite lament. Modern urban environments are often so chaotic, soulless, or ugly that I'm grateful for the humanizing touch of electronics. But I want to be aware of technology's effects, positive and negative. For music to remain vital, recordings have to exist in balance with live performance, and, these days, live performance is by far the smaller part of the equation. Perhaps we tell ourselves that we listen to CDs in order to get to know the music better or to supplement what we get from concerts and shows. But, honestly, a lot of us don't go to hear live music that often. Work leaves us depleted. Tickets are too expensive. Concert halls are stultifying. Rock clubs are full of kids who make us feel ancient. It's just so much easier to curl up in the comfy chair with a Beethoven quartet or Billie Holiday. But would Beethoven or Billie ever have existed if people had always listened to music the way we listen now?

"The machine is neither a god nor a devil," the German music critic Hans Stuckenschmidt wrote in 1926, in an essay on the mechanization of music. That eminently reasonable sentiment appears as an epigraph of Mark Katz's *Capturing Sound: How Technology Has Changed Music.* It's one of a number of recent books on the history of recording; two others are Colin Symes's *Setting the Record Straight: A Material History of Classical Recording,* which analyzes how the discourse around LPs and CDs shapes what we hear; and Robert Philip's *Performing Music in the Age of Recording,* which advances a potent thesis about how the phonograph transformed classical culture. Katz's book is the most approachable of these tomes. In lucid, evenhanded prose, it ranges all over the map, from classical to hip-hop. Although Katz believes that machines have profoundly affected how music is played and heard, he discourages a monolithic, deterministic idea of their impact. Ultimately, he says, the technology reflects whatever musical culture is exploiting it. The machine is a mirror of our needs and fears.

The principal irony of phonograph history is that the machine was not invented with music in mind. Edison conceived of his cylinder as a tool for business communication: it would replace the costly, imperfect practice of stenography and would have the added virtue of preserving in perpetuity the voices of the deceased. In an 1878 essay, Edison (or his ghostwriter) proclaimed portentously that his invention would "annihilate time and space, and bottle up for posterity the mere utterance of man." Annihilation is, of course, an ambiguous figure of speech. Recording broke down barriers between cultures, but it also placed more archaic musical forms in danger of extinction. In the early years of the century, Béla Bartók, Zoltán Kodály, and Percy Grainger used phonographs to preserve the voices of elderly folksingers whose timeless ways were being stamped out by the advance of modern life. And what was helping to stamp them out? The phonograph, with its international hit tunes and standardized popular dances.

In the 1890s, alert entrepreneurs installed phonographs in penny arcades, allowing customers to listen to favorite songs over ear tubes. By 1900, the phonograph was being marketed as a purely musical device. Its first great star was an operatic tenor, Enrico Caruso, whose voice remains one of the most transfixing phenomena in the history of the medium. The ping in his tone, that golden bark, penetrated the haze of the early technology and made the man himself viscerally present. Not so lucky was Johannes Brahms, who, in 1889, attempted to play his First Hungarian Dance for Edison's cylinder. It sounds as if the master were coming to us from a spacecraft disintegrating near Pluto. There was something symbolic in Edison's inability to register so titanic a presence as Brahms: despite Caruso's fame, and despite later fads for Toscanini, Bernstein, and Glenn Gould, classical music had a hard time getting a foothold in this slippery terrain. From the start, the phonograph favored brassy singing, knife-edged winds and brass, the thump of percussion—whatever could best puncture surface noise. Louis Armstrong's trumpet blasted through the crackle and pop of early records like no other instrument or voice of the time—he was Caruso's heir. Pianos, by contrast, were muddled and muffled; violins were all but inaudible. Classical music, with its softer-edged sounds, entered the recording era at a disadvantage. The age of the cowbell had begun.

Whenever a new gadget comes along, salespeople inevitably point out that an older gadget has been rendered obsolete. The automobile pushed aside the railroad; the computer replaced the typewriter. Sousa feared that the phonograph would supplant live music making. His fears were excessive but not irrational. Early ads for the phonograph took aim at the piano, which, around the turn of the century, was the center of domestic musical life, from the salon to the tavern. The top-selling Victrola of 1906 was encased in "piano-finished" mahogany, if anyone was missing the point. An ad reproduced in Colin Symes's book shows a family clustered about a phonograph, no piano in sight. Countless ad campaigns since have claimed that recordings are just as good as live performances, possibly better—combining, supposedly, the warmth of live music with the comfort of home. They have provided, to use some well-worn phrases, "the best seat in the house," "living presence," "perfect sound forever." They inspired the famous question, "Is it live or is it Memorex?" (If it's Memorex, is it dead?) Edison was so determined to demonstrate the verisimilitude of his machines that he held a nationwide series of Tone Tests, during which halls were plunged into darkness and audiences were supposedly unable to tell the difference between Anna Case singing live and one of her records.

It's easy to laugh now at the spectacle of the Tone Tests. Either Edison was engaging in serious hanky-panky, or audiences were so eager to embrace the new technology that they hypnotized themselves into ignoring the wheeze of cylinder static. But a hipper form of the same mumbo-jumbo is heard in high-end audio showrooms, where $10,000 systems purport to recreate an orchestra in your living room. Even if such a machine existed, the question once posed by the comedians Flanders and Swann lingers: Why would we want an orchestra in our living room? Isn't the idea of sitting in a room listening to a tape of 500 people performing the Mahler Eighth Symphony totally bizarre—the diametrical opposite of the great communal ceremonies that Mahler yearned to enact? So says the party of doom. The party of hope responds: Audiences generally ignored or misunderstood Mahler until repeated listening on LPs made his music comprehensible.

Like Heisenberg's mythical observer, the phonograph was never a mere recorder of events: it changed how people sang and played. Katz, in a major contribution to the lingo, calls these changes "phonograph effects." (The phrase comes from the digital studio, where it is used to describe the crackling, scratching noises that are sometimes added to pop-music tracks to lend them an appealingly antique air.) Katz devotes one striking chapter to a fundamental change in violin technique that took place in the early 20th century. It involved vibrato—that trembling action of the hand on the fingerboard, whereby the player is able to give notes a warbling sweetness. Until about 1920, vibrato was applied quite sparingly. On a 1903 recording, the great violinist Joseph Joachim uses it only to accentuate certain highly expressive notes. (The track is included on a CD that comes with Katz's book.) Around the same time, Fritz Kreisler began applying vibrato almost constantly. By the 1920s, most leading violinists had adopted Kreisler's method. Was it because they were imitating him? Katz proposes that the change came about for a more pedestrian reason. When a wobble was added to violin tone, the phonograph was able to pick it up more easily: it's a "wider" sound in acoustical terms, a blob of several superimposed frequencies. Also, the fuzzy focus of vibrato enabled players to cover up slight inaccuracies of intonation, and, from the start, the phonograph made players self-conscious about intonation in ways they had never been before. What worked in the studio then spread to the concert stage. Katz can't prove that the phonograph was responsible for the change, but he makes a good case.

Composers, who had reigned like gods over the dearly departed 19th century, were uncertain and quizzical in the face of the new device. Symes amusingly tracks the ambivalence of Igor Stravinsky, who styled himself the most impeccably up-to-date of composers. In 1916, the conductor Ernest Ansermet brought Stravinsky a stack of American pop records and sheet music, Jelly Roll Morton's "Jelly Roll Blues" possibly among them, and the composer swooned. "The musical ideal," he called them, "music spontaneous and 'useless,' music that wishes to express nothing." (Not quite what Jelly Roll had in mind.) Stravinsky began writing with the limitations of the phonograph in mind: short movements, small groups of instruments, lots of winds and brass, few strings. On his first American tour, in 1925, he signed a contract at Brunswick Studios, where Duke Ellington later set down "East St. Louis Toodle-O." Then, in the next decade, he abruptly adopted the John Philip Sousa line: "Oversaturated with sounds, blasé even before combinations of the utmost variety, listeners fall into a kind of torpor which deprives them of all power of discrimination." By the 1940s, Stravinsky was living in America, and, seeking new avenues of exposure, he embraced recording once again. He went so far as to endorse the Stromberg-Carlson Custom 400 loudspeaker, comparing it to a "fine microscope." You could try to find some consistent theory behind these statements, but the short version is that Stravinsky was confused.

The youngest composers of the 1920s—those who had come of age during and after the First World War—had no hesitation about submitting to the phonograph. Perhaps Katz's most fascinating chapter is devoted to the short-lived Grammophonmusik phenomenon in German music of the 1920s and early 1930s. Paul Hindemith, Kurt Weill, Ernst Toch, and Stefan Wolpe seized upon the phonograph not merely as a means for preserving and distributing music but as a way of making it. Wolpe was the first to take the plunge; at a Dada concert in 1920, he put eight phonographs on a stage and had them play parts of Beethoven's Fifth at different speeds. Weill wrote an interlude for solo record player—playing "Tango Angèle," his first "hit"—in the au-courant 1927 opera *The Tsar Has Himself Photographed.* Hindemith and Toch experimented with performances involving phonographs; fragmentary evidence of their legendary 1930 Gramophone Concert can be found on Katz's CD, and it's some of the craziest damn stuff you'll ever hear. We are only a step or two away from the electronic avant-garde of John Cage, whose "Imaginary Landscape No. 1," for piano, cymbals, and variable-speed turntables, dates from 1939. It turns out that the teenage Cage attended the Gramophone Concert during a summer break from school.

With the arrival of magnetic tape, the relationship between performer and medium became ever more complex. German engineers perfected the magnetic tape recorder, or Magnetophon, during the Second World War. Late one night, an audio expert turned serviceman named Jack Mullin was monitoring German radio when he noticed that an overnight orchestral broadcast was astonishingly clear: it sounded "live," yet not even at Hitler's whim could the orchestra have been playing Bruckner in the middle of the night. After the war was over, Mullin tracked down a Magnetophon and brought it to America. He demonstrated it to Bing Crosby, who used it to tape his broadcasts in advance. Crosby was a pioneer of perhaps the most famous of all technological effects, the croon. Magnetic tape meant that Bing could practically whisper into the microphone and still be heard across America; a marked drop-off in surface noise meant that vocal murmurs could register as vividly as Louis Armstrong's pealing trumpet.

Magnetic tape also meant that performers could invent their own reality in the studio. Errors could be corrected by splicing together bits of different takes. In the 1960s, the Beatles and the Beach Boys, following in the wake of electronic compositions by Cage and Stockhausen, began constructing intricate studio soundscapes that they never could have replicated onstage; even Glenn Gould would have had trouble executing the mechanically accelerated keyboard solo in "In My Life." The great rock debate about authenticity began. Were the Beatles pushing the art forward by reinventing it in the studio? Or were they losing touch with the earthy intelligence of folk, blues, and rock traditions? Bob Dylan stood at a craggy opposite extreme, turning out records in a few days' time and avoiding any vocal overdubs until *Blood on the Tracks,* the 14th record of his career. Yet frills-free, "lo-fi" recording has no special claim on musical truth; indeed, it easily becomes another phonograph effect, the effect of no effect. Even Dylan cannot escape the fictions of the medium, as he well knows: "I'm gazing out the window / Of the St. James Hotel / And I know no one can sing the blues / Like Blind Willie McTell."

In the 1980s, as Dutch and Japanese engineers introduced digital recording in the CD format, the saga of the phonograph experienced a final twist. Katz, in the last chapters of his book, delights in following the winding path from

Germany in the 1920s to the South Bronx in the 1970s, where the turntable became an instrument once again. DJs like Kool Herc, Afrika Bambaataa, and Grandmaster Flash used turntables to create a hurtling collage of phonograph effects—loops, breaks, beats, scratches. The silently observing machine was shoved into the middle of the party. It was assumed at first that this recording-driven music could never be recorded itself: the art of the DJ was all about fast moves over long duration, stamina and virtuosity combined. As Jeff Chang notes in his new book *Can't Stop Won't Stop: A History of the Hip-Hop Generation,* serious young d.j.s like Chuck D, on Long Island, laughed when a resourceful record company put out a rap novelty single called *Rapper's Delight.* How could a single record do justice to those endless parties in the Bronx where, in a multimedia rage of beats, tunes, raps, dances, and spray-painted images, kids managed to forget for a while that their neighborhood had become a smoldering ruin? The record labels found a way, of course, and a monster industry was born. Nowadays, hiphop fans are apt to claim that live shows are dead experiences, messy reenactments of pristine studio creations.

Recording has the unsettling power to transform any kind of music, no matter how unruly or how sublime, into a collectible object, which becomes decor for the lonely modern soul. It thrives on the buzz of the new, but it also breeds nostalgia, a state of melancholy remembrance and, with that, indifference to the present; you can start to feel nostalgic for the opening riff of a new favorite song even before you reach the end. Thomas Mann described the phonograph's ambiguous enchantments in the "Fullness of Harmony" chapter of *The Magic Mountain,* published in 1924. When a deluxe gramophone arrives at the Berghof sanitarium, it sends mixed messages to the young man who operates it. At times it sings "a new word of love" (shades of

Robert Johnson's "Phonograph Blues"); at times it exudes "sympathy for death." At the end of the novel, the hero goes marching toward an inferno of trench warfare, obliviously chanting the Schubert tune that the gramophone taught him. These days, he'd be rapping.

Throughout the 20th century, classical musicians and listeners together indulged the fantasy that they were living outside the technological realm. They cultivated an atmosphere of timelessness, of detachment from the ordinary world. Perhaps it's no accident that concert dress stopped evolving right about the time that Edison's cylinder came in: performers wished to prolong forever those last golden hours of the aristocratic age. Recording was well liked for its revenue-generating potential, but musicians preferred to think of it as a means of transcribing in the most literal manner the centuries-old classical performance tradition. With scattered exceptions—Weimar-era experimenters, postwar electronic composers, mavericks like Glenn Gould and the producer John Culshaw—musicians avoided the hey-let's-try-this spirit that defined pop recording from the start. As Symes points out, classical releases were prized for their unadorned realism. Recordings were supposed to deny the fact that they were recordings. That process involved, paradoxically, considerable artifice. Overdubbing, patching, knob-twiddling, and even digital effects such as "pitch correction" are as common in the classical studio as in pop. The phenomenon of the dummy star, who has a hard time replicating onstage what he or she purports to do on record, is not unheard of.

Robert Philip, in *Performing Music in the Age of Recording,* points out that the vaunted transparency of classical recording is often a micromanaged illusion and then goes further; he suggests that technology fundamentally altered the tradition that it was intended to preserve. Violin vibrato, as discussed in Mark Katz's book, is but one example of a phonograph effect in classical performance. Philip shows how every instrument in the orchestra acquired a standard profile. Listening to records became a kind of mirror stage through which musicians confronted their "true" selves. "Musicians who first heard their own recordings in the early years of the twentieth century were often taken aback by what they heard, suddenly being made aware of inaccuracies and mannerisms they had not suspected," Philip writes. As they adjusted their playing, they entered into a complex process that Katz calls a "feedback loop."

Feedback is what happens when an electric-guitar player gets too close to an amp and the amp starts squealing. Feedback in classical performance is the sound of musicians desperately trying to embody the superior self they glimpsed in the mirror and, potentially, turning themselves into robots in the process. Philip begins his book with a riveting description of concerts at the turn of the last century. "Freedom from disaster was the standard for a good concert," he writes. Rehearsals were brief, mishaps routine. Precision was not a universal value. Pianists rolled chords instead of playing them at one stroke. String players slid expressively from one note to the next—portamento, the style was called—in imitation of the slide of the voice. And the instruments themselves sounded different, depending on the nationality of the player. French bassoons had a reedy, pungent tone, quite unlike the rounded timbre of German bassoons. French flutists, by contrast, used more vibrato than their German and English counterparts, creating a warmer, mellower aura. American orchestral culture, which brought together immigrant musicians from all countries, began to erode the differences, and recordings canonized the emergent standard practice. Whatever style sounded cleanest on the medium—in these cases, German bassoons and French flutes—became the gold standard that players in conservatories copied. Young virtuosos today may have recognizable idiosyncrasies, but their playing seldom indicates that they came from any particular place or emerged from any particular tradition.

Archival reissues give tantalizing glimpses of the world as it was. Philip notes that in a 1912 performance the great Belgian violinist Eugène Ysaÿe "sways either side of the beat, while the piano maintains an even rhythm." In disks by the Bohemian Quartet, he says, "each player is functioning as an individual," reacting with seeming spontaneity to the personalities of the others. Edward Elgar's recordings of his Second Symphony and Cello Concerto, from 1927 and 1928, respectively, are practically explosive in impact, destroying all stereotypes of the composer as a staid Victorian gentleman. No modern orchestra would dare to play as the Londoners played for Elgar: phrases precipitously step over one another, tempos constantly change underfoot, rough attacks punch the clean surface. The biographical evidence suggests that this borderline-chaotic style of performance was exactly what Elgar wanted. "All sorts of things which other conductors carefully foster, he seems to leave to take their chance," a critic observed. Modern recordings of Elgar are so different in sound and spirit that they seem to document a different kind of music altogether. The symphonies have turned into monumental processional rituals, along the lines of the symphonies of Bruckner or at least the version of Bruckner that conductors now give us.

All those lost tics and traits—swaying on either side of the beat, sliding between notes, breaking chords into arpeggios, members of a quartet going every which way—are alike in bringing out the distinct voices of the players, not to mention the mere fact that they are fallible humans. Philip writes, "If you hear the Royal Albert Hall Orchestra sliding, you may or may not like it, but you cannot be unaware of the physical process of playing." Most modern performance tends to erase all evidence of the work that goes into playing: virtuosity is defined as effortlessness. One often-quoted ideal is to "disappear behind the music." But when precision is divorced from emotion it can become antimusical, inhuman, repulsive.

Is there any escape from the "feedback loop"? Philip, having blamed recordings for a multitude of sins, ends by saying that they might be able to come to the rescue. By studying artifacts from the dawn of the century, musicians might recapture what has gone missing from the perfectionist style. They can rebel against the letter of the score in pursuit of its spirit. But there are enormous psychic barriers in the way of such a shift: performers will have to be unafraid of indulging mannerisms that will sound sloppy to some ears, of committing what will sound like mistakes. They will have to defy the hypercompetitive conservatory culture in which they came of age, and also the hyperprofessionalized culture of the ensembles in which they find work.

In at least one area, though, performance style has undergone a sea change. Early music has long had the reputation of being the most pedantically "correct" subculture in classical music; Philip exposes its contradictions in one chapter of his book. But the more dynamic Renaissance and baroque specialists—Jordi Savall, Andrew Manze, the Venice Baroque Orchestra, Il Giardino Armonico, William Christie's Les Arts Florissants—are exercising all the freedoms that Philip misses in modern performance: they execute some notes cleanly and others roughly; they weave around the beat instead of staying right on top of it; they slide from note to note when they are so moved. As a result, the music feels liberated, and audiences tend to respond in kind, with yelps of joy.

Philip, at the end of his masterly thesis, is left with an uncertainty. No matter how much evidence he accumulates, he can't quite prove that classical playing became standardized because the phonograph demanded it. Records cannot be entirely to blame, he admits: otherwise, similar patterns would surface in popular music, which, whatever its problems, has never lacked for spontaneity. The urge toward precision was already well under way in the late 19th century, when Hans von Bülow's Meiningen orchestra was celebrated as the best-rehearsed of its time and when the big new orchestras of America, the Boston Symphony first and foremost, astonished European visitors like Richard Strauss and Gustav Mahler with the discipline of their playing. Other technologies that preceded the phonograph also changed how people played and listened. Those who got to know music on a well-tuned piano began to expect the same from an orchestra. The sonic wonders of Boston's Symphony Hall—the first hall whose acoustics were scientifically designed—placed a golden frame around the music, and the orchestra had to measure up. Most of all, classical music in America suffered from being a reproduction itself, an immaculate copy of European tradition. We've been listening to the same record for a century and a half.

Twenty years ago, the American composer Benjamin Boretz wrote, "In music, as in everything, the disappearing moment of experience is the firmest reality." The paradox of recording is that it can preserve forever those disappearing moments of sound but never the spark of humanity that generates them. This is a paradox common to technological existence: everything gets a little easier and a little less real.

Then again, the reigning unreality of the electronic sphere can set us up for a new kind of ecstasy, once we unplug ourselves from our gadgets and expose ourselves to the risk of live performance. Recently at Carnegie Hall, Gidon Kremer and the Baltimore Symphony played Shostakovich's First Violin Concerto, and over and above the physical power of Kremer's playing—his tone ran the gamut from the gnawingly raw to the angelically pure—the performance offered the shock of the real: on an average, bustling New York night, Shostakovich bore down on the audience like a phantom train.

In 1964, Glenn Gould made a famous decision to renounce live performance. In an essay published two years later, "The Prospects of Recording," he predicted that the concert would eventually die out, to be replaced by a purely electronic music culture. He may still be proved right. For now, live performance clings to life and, in tandem, the classical-music tradition that could hardly exist without it. As the years go by, Gould's line of argument, which served to explain his decision to abandon the concert stage, seems ever more misguided and dangerous. Gould praised recordings for their vast archival possibilities, for their ability to supply on demand a bassoon sonata by Hindemith or a motet by Buxtehude. He gloried in the extraordinary interpretive control that studio conditions allowed him. He took it for granted that the taste for Buxtehude motets or for surprising new approaches to Bach could survive the death of the concert—that somehow new electronic avenues could be found to spread the word about old and unusual music. Gould's thesis is annulled by cold statistics: classical-record sales have plunged, while concert attendance is anxiously holding steady. Ironically, Gould himself remains, posthumously, one of the last blockbuster classical recording artists: Sony Classical's recent rerelease of his two interpretations of

Bach's Goldberg Variations sold 200,000 copies. That's surely not what Gould had in mind for the future of the medium.

A few months after Gould published his essay, the Beatles, in a presumably unrelated development, played their last live show, in San Francisco. They spent the rest of their short career working in the recording studio. They proved, as did Gould, that the studio breeds startlingly original ideas; they also proved, as did Gould, that it breeds a certain kind of madness. I'll take *Rubber Soul* over *Sgt. Pepper's,* and Gould's 1955 Goldbergs over his 1981 version, because the first recording in each pair is the more robust, the more generous, the more casually sublime. The fact that the Beatles broke up three years after they disappeared into the studio, and the fact that Gould died in strange psychic shape at the age of 50, may tell us all we need to know about the seductions and sorrows of the art of recording.

David Bernstein

When the Sous-Chef Is an Inkjet

A glimpse at the future of food

Homaro Cantu's maki look a lot like the sushi rolls served at other upscale restaurants: pristine, coin-size disks stuffed with lumps of fresh crab and rice and wrapped in shiny nori. They also taste like sushi, deliciously fishy and seaweedy.

But the sushi made by Mr. Cantu, the 28-year-old executive chef at Moto in Chicago, often contains no fish. It is prepared on a Canon i560 inkjet printer rather than a cutting board. He prints images of maki on pieces of edible paper made of soybeans and cornstarch, using organic, food-based inks of his own concoction. He then flavors the back of the paper, which is ordinarily used to put images onto birthday cakes, with powdered soy and seaweed seasonings.

At least two or three food items made of paper are likely to be included in a meal at Moto, which might include 10 or more tasting courses. Even the menu is edible; diners crunch it up into a bowl of gazpacho, creating Mr. Cantu's version of alphabet soup.

Sometimes he seasons the menus to taste like the main courses. Recently, he used dehydrated squash and sour cream powders to match a soup entrée. He also prepares edible photographs flavored to fit a theme: an image of a cow, for example, might taste like filet mignon.

"We can create any sort of flavor on a printed image that we set our minds to," Mr. Cantu said. The connections need not stop with things ordinarily thought of as food. "What does M. C. Escher's *Relativity* painting taste like? That's where we go next."

Food critics have cheered, comparing Mr. Cantu to Salvador Dali and Willy Wonka for his peculiarly playful style of cooking. More precisely, he is a chef in the Buck Rogers tradition, blazing a trail to a space-age culinary frontier.

Mr. Cantu wants to use technology to change the way people perceive (and eat) food, and he uses Moto as his laboratory. "Gastronomy has to catch up to the evolution in technology," he said. "And we're helping that process happen."

Tucked among warehouses and lofts in the Chicago meatpacking district, Moto attracts a trend-conscious crowd. Some guests leave scratching their heads; others walk away spellbound by a glimpse of Mr. Cantu's vision of the future of food.

William Mericle, 41, described a recent meal at Moto as "dinner theater on your plate." He did not care for all 20 small dishes he sampled, but he said he liked most of them. He found Mr. Cantu's imagination appealing. "He's mad-scientist-meets-gourmet-chef," he said. "Like Christopher Lloyd from *Back to the Future,* if he were more interested in food than time travel."

Mr. Cantu believes that restaurant-goers, particularly diners who are willing to spend $240 per person for a meal (the cost of a 20-course tasting menu with wine at Moto), are often disappointed by conventional dining experiences. "They're sick and tired of steak and eggs," he said. "They're tired of just going to a restaurant, having food placed on the table, having it cleared, and there's no more mental input into it other than the basic needs of a caveman, just eat and nourish."

At Moto, he said, "there's so much more we can do."

Mr. Cantu is experimenting with liquid nitrogen, helium, and superconductors to make foods levitate. And while many chefs speak of buying new ovens or refrigerators, he wants to invest in a three-dimensional printer to make physical prototypes of his inventions, which he now painstakingly builds by hand. The 3-D printer could function as a cooking device, creating silicone molds for pill-sized dishes flavored, say, like watermelon, bacon and eggs, or even beef Bourguignonne, he said, and he could also make edible molds out of cornstarch.

He also plans to buy a class IV laser to create dishes that are "impossible through conventional means." (A class IV laser, the highest grade under the Occupational Safety and Health Administration's classification system, projects high-powered beams and is typically used for surgery or welding.)

Mr. Cantu said he might use the laser to burn a hole through a piece of sashimi tuna, cooking the fish thoroughly inside but leaving its exterior raw. He said he would also use the laser to create "inside out" bread, where the crust is baked inside the loaf and the doughy part is the outer surface. "We'll be the first restaurant on planet Earth to use a class IV laser to cook food," he said with a grin.

He is testing a handheld ion-particle gun, which he said is for levitating food. So far he has zapped only salt and sugar but envisions one day making whole meals float before awestruck diners.

The son of a fabricating engineer, Mr. Cantu got his start as a science geek. "From a very young age, I liked to take apart things," said Mr. Cantu, who grew up in the Pacific Northwest. "All of my Christmas gifts would wind up in a million pieces. I actually recall taking apart my dad's lawn mower three times to understand how combustible engines work."

When he was 12, he took a job as a cook and busboy, mainly to earn money for remote-controlled airplanes and helicopters that he then took apart. But the restaurant business rubbed off on Mr. Cantu, and after high school he attended culinary school at Le Cordon Bleu in Portland, Oregon. A series of jobs followed, nearly 50 in all, Mr. Cantu said. He worked as a stagiaire, or intern, in some of the top kitchens around the country, eventually talking his way into a job at Charlie Trotter's, a well-known restaurant in Chicago. He became a sous-chef there before opening Moto last year.

Mr. Cantu has filed applications for patents on more than 30 inventions, including a cooking box that steams fish. The tiny opaque box, about three inches square, is made of a superinsulating polymer. Mr. Cantu heats the box to 350 degrees in an oven and places a raw piece of Pacific sea bass inside it. A server then delivers it to diners, who can watch the fish cook.

Assisting Mr. Cantu with what he calls his "'Star Wars' stuff" is DeepLabs, a small Chicago product-development and design consultancy. Mr. Cantu meets weekly with the crew of aerospace and mechanical engineers, programmers, and product designers at DeepLabs for brainstorming sessions.

"I tell them I want to make food float; I want to make it disappear; I want to make it reappear; I want to make the utensils edible; I want to make the plates, the table, the chairs edible," Mr. Cantu said. "I ask them, what do I need to do that?"

Ryan Alexander, an industrial graphic designer at DeepLabs, said he and his colleagues at the company, which has designed more conventional products for Motorola and Home Depot, are enthusiastic about Mr. Cantu: "We don't say no," he said.

Using engineering, graphics, and animation software, DeepLabs designers have begun to turn Mr. Cantu's dreams into realities.

They have created mock-ups of his all-in-one utensil, a combination fork, knife, and spoon, as well as utensils with pressurized handles that release aromatic vapors. The latest prototype is a utensil with a disposable, self-heating silicone handle that can be filled with liquefied or pureed foods. A carbon-dioxide-based charge heats the food (soup, for example), and the diner squeezes the handle to release it onto a spoon. Mr. Cantu envisions many applications for such a utensil, from military meals to cookouts.

Mr. Cantu said his experiments and kitchen inventions could one day revolutionize how, where, and what we eat. "This will tap into something," he said. "Maybe a mission to Mars; I don't know. Maybe we're going to find a way to grow something in a temperature that liquid nitrogen operates at. Then we could grow food on Pluto. There are possibilities to this that we can't fathom yet. And to not do it is far more consequential than just to say, hey, we're going to stick with our steak and eggs today."

Edward Tenner

The Rise of the Plagiosphere

How new tools to detect plagiarism could induce writer's block.

The 1960s gave us, among other mind-altering ideas, a revolutionary new metaphor for our physical and chemical surroundings: the biosphere. But an even more momentous change is coming. Emerging technologies are causing a shift in our mental ecology, one that will turn our culture into the plagiosphere, a closing frontier of ideas.

The Apollo missions' photographs of Earth as a blue sphere helped win millions of people to the environmentalist view of the planet as a fragile and interdependent whole. The Russian geoscientist Vladimir Vernadsky had coined the word *biosphere* as early as 1926, and the Yale University biologist G. Evelyn Hutchinson had expanded on the theme of Earth as a system maintaining its own equilibrium. But as the German environmental scholar Wolfgang Sachs observed, our imaging systems also helped create a vision of the planet's surface as an object of rationalized control and management—a corporate and unromantic conclusion to humanity's voyages of discovery.

What NASA did to our conception of the planet, Web-based technologies are beginning to do to our understanding of our written thoughts. We look at our ideas with less wonder and with a greater sense that others have already noted what we're seeing for the first time. The plagiosphere is arising from three movements: Web indexing, text matching, and paraphrase detection.

The first of these movements began with the invention of programs called Web crawlers, or spiders. Since the mid-1990s, they have been perusing the now billions of pages of Web content; indexing every significant word found; and making it possible for Web users to retrieve without charge and in fractions of seconds, pages with desired words and phrases.

The spiders' reach makes searching more efficient than most of technology's wildest prophets imagined, but it can yield unwanted knowledge. The clever phrase a writer coins usually turns out to have been used for years worldwide—used in good faith because until recently the only way to investigate priority was in a few books of quotations. And in our accelerated age, even true uniqueness has been limited to 15 minutes. Bons mots that once could have enjoyed a half-life of a season can decay overnight into clichés.

Still, the major search engines have their limits. Alone, they can check a phrase, perhaps a sentence, but not an extended document. And at least in their free versions, they generally do not produce results from proprietary databases like LexisNexis, Factiva, ProQuest, and other paid-subscription sites or from free databases that dynamically generate pages only when a user submits a query. They also don't include most documents circulating as electronic manuscripts with no permanent Web address.

Enter text-comparison software. A small handful of entrepreneurs have developed programs that search the open Web and proprietary databases, as well as e-books, for suspicious matches. One of the most popular of these is Turnitin; inspired by journalism scandals such as the *New York Times*'s Jayson Blair case, its creators offer a version aimed at newspaper editors. Teachers can submit student papers electronically for comparison with these databases, including the retained texts of previously submitted papers. Those passages that bear resemblance to each other are noted with color highlighting in a double-pane view.

Two years ago I heard a speech by a New Jersey electronic librarian who had become an antiplagiarism specialist and consultant. He observed that comparison programs were so thorough that they often flagged chance similarities between student papers and other documents. Consider, then, that Turnitin's spiders are adding 40 million pages from the public Web, plus 40,000 student papers, each day. Meanwhile Google plans to scan millions of library books—including many still under copyright—for its Print database. The number of coincidental parallelisms between the various things that people write is bound to rise steadily.

A third technology will add yet more capacity to find similarities in writing. Artificial-intelligence researchers at MIT and other universities are developing techniques for identifying nonverbatim similarity between documents to make possible the detection of nonverbatim plagiarism. While the investigators may have in mind only cases of brazen paraphrase, a program of this kind can multiply the number of parallel passages severalfold.

Some universities are encouraging students to precheck their papers and drafts against the emerging plagiosphere. Perhaps publications will soon routinely screen submissions. The problem here is that while such rigorous and robust policing will no doubt reduce cheating, it may also give writers a sense of futility. The concept of the biosphere exposed our environmental fragility; the emergence of the plagiosphere perhaps represents our textual impasse. Copernicus may have deprived us of our centrality in the cosmos and Darwin of our uniqueness in the biosphere, but at least they left us the illusion of the originality of our words. Soon that, too, will be gone.

Clive Thompson

The Xbox Auteurs

How Michael Burns and his fellow tech-savvy cineastes are making movies set entirely inside video games

Like many young hipsters in Austin, Texas, Michael Burns wanted to make it big in some creative field—perhaps writing comedy scripts in Hollywood. Instead, he wound up in a dead-end job, managing a call center. To kill time, he made friends with a group of equally clever and bored young men at the company where he worked, and they'd sit around talking about their shared passion: video games. Their favorite title was Halo, a best-selling Xbox game in which players control armor-clad soldiers as they wander through gorgeous coastal forests and grim military bunkers and fight an army of lizardlike aliens. Burns and his gang especially loved the "team versus team" mode, which is like a digital version of paintball: instead of fighting aliens, players hook their Xboxes to the Internet and then log on together in a single game, at which point they assemble into two teams—red-armored soldiers versus blue-armored ones. Instead of shooting aliens, they try to slaughter one another, using grenades, machine guns, and death rays. On evenings and weekends, Burns and his friends would cluster around their

TVs until the wee hours of the morning, gleefully blowing one another to pieces.

"Halo is like crack," Burns recalls thinking. "I could play it until I die."

Whenever a friend discovered a particularly cool stunt inside Halo—for example, obliterating an enemy with a new type of grenade toss—Burns would record a video of the stunt for posterity. (His friend would perform the move after Burns had run a video cord from his TV to his computer, so he could save it onto his hard drive.) Then he'd post the video on a Web site to show other gamers how the trick was done. To make the videos funnier, sometimes Burns would pull out a microphone and record a comedic voice-over, using video-editing software to make it appear as if the helmeted soldier himself were doing the talking.

Then one day he realized that the videos he was making were essentially computer-animated movies, almost like miniature emulations of *Finding Nemo* or *The Incredibles.* He was using the game to function like a personal Pixar studio. He wondered: Could he use it to create an actual movie or TV series?

Burns's group decided to give it a shot. They gathered around the Xbox at Burns's apartment, manipulating their soldiers like tiny virtual actors, bobbing their heads to look as if they were deep in conversation. Burns wrote sharp, sardonic scripts for them to perform. He created a comedy series called *Red vs. Blue,* a sort of sci-fi version of M*A*S*H. In *Red vs. Blue,* the soldiers rarely do any fighting; they just stand around insulting one another and musing over the absurdities of war, sounding less like patriotic warriors than like bored, clever, video-store clerks. The first 10-minute episode opened with a scene set in Halo's bleakest desert canyon. Two red soldiers stood on their base, peering at two blue soldiers far off in the distance, and traded quips that sounded almost like a slacker disquisition on Iraq:

> *Red Soldier:* "Why are we out here? Far as I can tell, it's just a box canyon in the middle of nowhere, with no way in or out. And the only reason we set up a red base here is because they have a blue base there. And the only reason they have a blue base over there is because we have a red base here."

When they were done, they posted the episode on their Web site (surreptitiously hosted on computers at work). They figured maybe a few hundred people would see it and get a chuckle or two.

Instead, *Red vs. Blue* became an instant runaway hit on geek blogs, and within a single day, twenty thousand people stampeded to the Web site to download the file. The avalanche of traffic crashed the company server. "My boss came into the office and was like, 'What the hell is going on?'" Burns recalls. "I looked over at the server, and it was going blink, blink, blink."

Thrilled, Burns and his crew quickly cranked out another video and then another. They kept up a weekly production schedule, and after a few months, *Red vs. Blue* had, like some dystopian version of *Friends,* become a piece of appointment viewing. Nearly a million people were downloading each episode every Friday, writing mash notes to the creators and asking if they could buy a DVD of the collected episodes. Mainstream media picked up on the phenomenon. The *Village Voice* described it as "'Clerks' meets 'Star Wars,'" and the BBC called it "riotously funny" and said it was "reminiscent of the anarchic energy of 'South Park.'" Burns realized something strange was going on. He and his crew had created a hit comedy show—entirely inside a video game.

Video games have not enjoyed good publicity lately. Hillary Clinton has been denouncing the violence in titles like Grand Theft Auto, which was yanked out of many stores recently amid news that players had unlocked sex scenes hidden inside. Yet when they're not bemoaning the virtual bloodshed, cultural pundits grudgingly admit that today's games have become impressively cinematic. It's not merely that the graphics are so good: the camera angles inside the games borrow literally from the visual language of film. When you're playing Halo and look up at the sun, you'll see a little "lens flare," as if you were viewing the whole experience through the eyepiece of a 16-millimeter Arriflex. By using the game to actually make cinema, Burns and his crew flipped a switch that neatly closed a self-referential media loop: movies begat games that begat movies.

And Burns and his crew aren't alone. Video-game aficionados have been creating *machinima*—an ungainly term mixing *machine* and *cinema* and pronounced ma-SHEEN-i-ma—since the late 1990s. *Red vs. Blue* is the first to break out of the underground, and now corporations like Volvo are hiring machinima artists to make short promotional films, while MTV, Spike TV, and the Independent Film Channel are running comedy shorts and music videos produced inside games. By last spring, Burns and his friends were making so much money from *Red vs. Blue* that they left their jobs and founded Rooster Teeth Productions. Now they produce machinima full-time.

It may be the most unlikely form of indie filmmaking yet—and one of the most weirdly democratic. "It's like 'The Blair Witch Project' all over again, except you don't even need a camera," says Julie Kanarowski, a product manager with Electronic Arts, the nation's largest video-game publisher. "You don't even need actors."

Back in college, Burns and another Rooster Teeth founder, Matt Hullum, wrote and produced a traditional live-action indie movie. It cost $9,000, required a full year to make, and was seen by virtually no one. By contrast, the four Xboxes needed to make *Red vs. Blue* cost a mere $600. Each 10-minute episode requires a single day to perform and edit and is viewed by hordes of feverish video-game fans the planet over.

More than just a cheap way to make an animated movie, machinima allows game players to comment directly on the pop culture they so devotedly consume. Much like "fan fiction" (homespun tales featuring popular TV characters) or "mash-ups" (music fans blending two songs to create a new hybrid), machinima is a fan-created art form. It's what you get when gamers stop blasting aliens for a second and start messing with the narrative.

And God knows, there's plenty to mess with. These days, the worlds inside games are so huge and open-ended that gamers can roam anywhere they wish. Indeed, players often abandon the official goal of the game—save the princess; vanquish the eldritch forces of evil—in favor of merely using the virtual environment as a gigantic jungle gym. In one popular piece of Halo machinima, "Warthog Jump," a player cunningly used the game to conduct a series of dazzling physics experiments. He placed grenades in precise locations beneath jeeps and troops, such that when the targets blew sky high, they pinwheeled through the air in precise formations, like synchronized divers. Another gamer recorded a machinima movie that poked subversive fun at Grand Theft Auto. Instead of playing as a dangerous, cop-killing gangster, the player pretended he was a naive

Canadian tourist—putting down his gun, dressing in tacky clothes, and simply wandering around the game's downtown environment for hours, admiring the scenery.

So what's it like to actually shoot a movie inside a game? In June, I visited the Rooster Teeth offices in Buda, Texas, a tiny Austin suburb, to observe Burns and his group as they produced a scene of *Red vs. Blue.* Burns, a tall, burly 32-year-old, sat in front of two huge flat-panel screens, preparing the editing software. Nearby were the two Rooster Teeth producers who would be acting on-screen: Geoff Ramsey, a scraggly-bearded 30-year-old whose arms are completely covered in tattoos of fish and skulls, and Gustavo Sorola, a gangly 27-year-old who sprawled in a beanbag chair and peered through his thick architect glasses at the day's e-mail. They were fan letters, Sorola told me, that pour in from teenagers who are as enthusiastic as they are incoherent. "The way kids write these days," he said with a grimace. "It's like someone threw up on the keyboard."

In the script they were acting out that day, a pair of *Red vs. Blue* soldiers engaged in one of their typically pointless existential arguments, bickering over whether it's possible to kill someone with a toy replica of a real weapon. The Rooster Teeth crew recorded the voice-overs earlier in the day; now they were going to create the animation for the scene.

Burns picked up a controller and booted up Halo on an Xbox. He would act as the camera: whatever his character saw would be recorded from his point of view. Then Sorola and Ramsey logged into the game, teleporting in as an orange-suited and a red-suited soldier. Burns posed them near a massive concrete bunker and frowned as he scrutinized the view on the computer screen. "Hmmmm," he muttered. "We need something to frame you guys—some sort of prop." He ran his character over to a nearby alien hovercraft, jumped in and parked it next to the actors. "Sweet!" he said. "I like it!"

In a *Red vs. Blue* shoot, the actors all must follow one important rule: Be careful not to accidentally kill another actor. "Sometimes you'll drop your controller and it unintentionally launches a grenade. It takes, like, 20 minutes for the blood splatters to dry up," Ramsey said. "Totally ruins the scene."

Finally, Burns was ready to go. He shouted, "Action!" and the voice-overs began playing over loudspeakers. Sorola and Ramsey acted in time with the dialogue. Acting, in this context, was weirdly minimalist. They mashed the controller joysticks with their thumbs, bobbing the soldiers' heads back and forth roughly in time with important words in each line. "It's puppetry, basically," Ramsey said, as he jiggled his controller. Of all the *Red vs. Blue* crew members, Ramsey is renowned for his dexterity with an Xbox. When a scene calls for more than five actors onstage, he'll put another controller on the ground and manipulate it with his right foot, allowing him to perform as two characters simultaneously.

As I watched, I was reminded of what initially cracked me up so much about *Red vs. Blue:* the idea that faceless, anonymous soldiers in a video game have interior lives. It's a "Rosencrantz and Guildenstern" conceit; *Red vs. Blue* is what the game characters talk about when we're not around to play with them. As it turns out, they're a bunch of neurotics straight out of *Seinfeld.* One recruit reveals that he chain-smokes inside his airtight armor; a sergeant tells a soldier his battle instructions are to "scream like a woman." And, in a sardonic gloss on the game's endless carnage, none of the soldiers have the vaguest clue why they're fighting.

Yet as I discovered, real-life soldiers are among the most

ardent fans of *Red vs. Blue.* When I walked around the Rooster Teeth office, I found it was festooned with letters, plaques, and an enormous American flag, gifts from grateful American troops, many of whom are currently stationed in Iraq. Isn't it a little astonishing, I asked Burns when the crew went out in the baking Texas sun for a break, that actual soldiers are so enamored of a show that portrays troops as inept cowards, leaders as cynical sociopaths, and war itself as a supremely meaningless endeavor? Burns laughed but said the appeal was nothing sinister.

"*Red vs. Blue* is about downtime," he said. "There's very little action, which is precisely the way things are in real life."

"He's right," Ramsey added. He himself spent five years in the army after high school. "We'd just sit around digging ditches and threatening to kill each other all day long," he said. "We were bored out of our minds."

Perhaps the most unusual thing about machinima is that none of its creators are in jail. After all, they're gleefully plundering intellectual property at a time when the copyright wars have become particularly vicious. Yet videogame companies have been upbeat—even exuberant—about the legions of teenagers and artists pillaging their games. This is particularly bewildering in the case of *Red vs. Blue,* because Halo is made by Bungie, a subsidiary of Microsoft, a company no stranger to using a courtroom to defend its goods. What the heck is going on?

As it turns out, people at Bungie love *Red vs. Blue.* "We thought it was kind of brilliant," says Brian Jarrard, the Bungie staff member who manages interactions with fans. "There are people out there who would never have heard about Halo without *Red vs. Blue.* It's getting an audience outside the hardcore gaming crowd."

Sure, Rooster Teeth ripped off Microsoft's intellectual

property. But Microsoft got something in return: *Red vs. Blue* gave the game a whiff of countercultural coolness, the sort of grassroots street cred that major corporations desperately crave but can never manufacture. After talking with Rooster Teeth, Microsoft agreed, remarkably, to let them use the game without paying any licensing fees at all. In fact, the company later hired Rooster Teeth to produce *Red vs. Blue* videos to play as advertisements in game stores. Microsoft has been so strangely solicitous that when it was developing the sequel to Halo last year, the designers actually inserted a special command—a joystick button that makes a soldier lower his weapon—designed solely to make it easier for Rooster Teeth to do dialogue.

"If you're playing the game, there's no reason to lower your weapon at all," Burns explained. "They put that in literally just so we can shoot machinima."

Other game companies have gone even further. Many now include editing software with their games, specifically to encourage fans to shoot movies. When Valve software released its hit game Half-Life 2 last year, it included "Faceposer" software so that machinima creators could tweak the facial expressions of characters. When the Sims 2—a sequel to the top-selling game of all time—came out last year, its publisher, Electronic Arts, set up a Web site so that fans could upload their Sims 2 movies to show to the world. (About 8,000 people so far have done so.)

Still, it's one thing for gamers to produce a jokey comedy or a music video. Can machinima actually produce a work of art—something with serious emotional depth? A few people have tried. In China, a visual artist named Feng Mengbo used the first-person-shooter game Quake III to produce Q4U, in which the screen is filled with multiple versions of himself, killing one another. Players' relationships with constant, blood-splattering violence are a common subject in game art. Last year, the 31-year-old artist Brody Condon produced an unsettling film that consisted of nothing but shots of himself committing suicide inside 50 different video games.

"I try to come to terms with what taking your life means in these games," Condon says. "I'm trying to understand, spiritually, your relationship with an avatar on the screen."

But even machinima's biggest fans admit that the vast majority of machinima is pretty amateurish. "It's like if some friends of mine all broke into a movie set, and we all got to use all the cameras and special-effects equipment," says Carl Goodman, director of digital media at the American Museum of the Moving Image, which began to hold an annual machinima festival two years ago. "We wouldn't quite know how to use it, but we'd make some pretty interesting stuff."

Yet as Goodman points out, there's a competing proposition. Machinima does not always strive to emulate "realistic," artistic movies. On the contrary, it is often explicitly devoted to celebrating the aesthetics of games—the animations and in-jokes, the precise physics. Most machinima is probably meaningless to people who don't play games, much as ESPN is opaque to anyone who doesn't watch sports. But for those who do play Halo, it was genuinely thrilling to see something like "Warthog Jump," with its meticulously synchronized explosions.

The Rooster Teeth crew has its own hilariously stringent rule for making machinima: no cheating. When they shoot *Red vs. Blue,* they do not use any special effects that are not organically included in the game; everything you see in an episode of *Red vs. Blue* could in theory have taken place during an actual game of Halo, played by a fan in his or her bedroom. It's a charmingly purist attitude, a sci-fi version of the "Dogma" school of indie film, which argues that movies are best when cinematic trickery is kept to a minimum.

One evening in New York, I visited with Ethan Vogt as he and his machinima team shot a car-chase scene for a Volvo promo. Vogt and two producers sat at computers, logged into a multiplayer game; each producer controlled a car racing through crowded city streets, while Vogt controlled a free-floating "camera" that followed behind, recording the visuals. The vehicles—an enormous 1972 Chevy Impala and a Volvo V50—screamed along at about 60 miles an hour, fishtailing through corners while plowing into mailboxes; lampposts; and, occasionally, clots of pedestrians. The lead car burst into flames. "That's great," Vogt said. "That's great."

Though it shares with independent filmmaking a do-it-yourself aesthetic, machinima inverts the central tradition of indie film: smallness. With their skimpy budgets, indie directors tend to set movies in kitchens or living rooms—and focus instead on providing quality acting and scripts. Machinima, in contrast, often has horribly cheesy acting and ham-fisted, purple-prose stories—but they're set in outer space. Want massive shootouts? Howling mob scenes? Roman gladiatorial armies clashing by night? No problem. It is the rare form of amateur film in which the directors aspire to be not Wes Anderson but George Lucas.

Indeed, with video games played on computers, it is now possible to build an entire world from scratch. The core of any video game is its game engine, the software that knows how to render 3-D objects and how to realistically represent the physics of how they move, bounce, or collide. But the actual objects inside the game—the people, the cars, the guns, even the buildings—can be altered, tweaked, or replaced by modifications, or "mods." Mods do not require any deep programming skills; indeed, almost any teenager with a passing acquaintance with graphic-design software can "re-skin" a character in a game to make it look like himself or herself, for instance. (Xbox and PlayStation games, in comparison, are much harder to mod, because the consoles are locked boxes, designed to prevent players from tampering with the games.)

I was able to see modding in action one night when I visited the ILL Clan, a pioneering machinima group. Their headquarters are the kitchen table in the cramped one-bedroom Brooklyn apartment of Frank Dellario; a lanky, hyperkinetic 42-year-old, he sat on a rickety folding chair, pecking at a keyboard. The table was littered with four computer screens and laptops, the remnants of take-out sushi, and a hopelessly tangled morass of computer cords and joysticks; a huge wide-screen TV lurked behind them for viewing their work. On the night I visited, they were using a game engine called Torque to shoot a short heist movie for Audi, in which two thugs beat up a concert violinist and make off with an antique violin in a van.

To quickly create a gritty-looking city, Dellario and his colleague—ILL Clan's cofounder, Matt Dominianni—hired a local artist to build a generic-looking urban intersection inside the game. To customize it, Dominianni went onto Google, found snapshots of a few seedy stores (an adult bookstore, a tattoo parlor, and a furniture outlet), and digitally pasted them onto the front of the buildings. Then they went to a site called Turbo-Squid, a sort of Amazon for virtual in-game items, and for $45 dollars bought a van that could be plunked down inside the game. When I arrived, they were browsing the site and contemplating buying a few women. "My God, look at this one," Dellario marveled, as he clicked open a picture of an eerily realistic 3-D brunette named Masha. "I'm going to marry this woman. They've finally broken through to total reality."

Dellario put the van into the correct location in the scene and then logged into the game to figure out the camera angle for this shot. He frowned. It didn't look right. The lighting was all off, with shadows falling in the wrong places.

Dominianni figured out the problem: "The sun is supposed to be at high noon. It's in the wrong place."

"Oh, yeah," Dellario said. "Let me move it." He pulled up a menu, clicked on the "sun" command, and dragged it across the sky.

Now they were finally ready to shoot. Dellario realized they needed an extra pair of hands to manipulate one of the thugs. "Want to act in this scene?" Dellario asked, and he handed me a joystick.

I sat down at one of the computers and took control of "Thug1," a brown-haired man in a golf shirt and brown pants, carrying the stolen violin. Dominianni was playing "Thug2." Our characters were supposed to look around to make sure the coast was clear and then jump in the truck and race off. Dellario gave me my motivation: "It's like you hear a suspicious noise. You're nervous." I used the joystick to practice moving my virtual character, craning its neck—my neck?—back and forth. I have played plenty of video games, but this felt awfully odd. Usually when I am inside a game, I'm just worried about staying alive while the bullets whiz past my ears. I've never had to emote.

While Dellario and Dominianni fiddled with the camera angle, I grew impatient and wandered around, exploring the virtual set. I peered in a few shop windows—they were strikingly photorealistic, even up close. Then I walked down an alley and suddenly arrived at the end of the set. It was like a tiny Western town in the desert: once you got beyond the few clustered buildings, there was nothing there—just a vast, enormous plain, utterly empty and stretching off infinitely into the distance.

This spring, Electronic Arts decided to promote the Sims 2 by hiring Rooster Teeth to create a machinima show using the game. Called *The Strangerhood,* it would be freely available online. *The Strangerhood* is a parody of reality TV: a group of people wake up one day to discover that they are living in new houses and they can't remember who they are or how they got there. In the Sims 2, the animated people are impressively Pixar-like and expressive, making *The Strangerhood* even more like a mainstream animated show than *Red vs. Blue;* you could almost imagine watching it on Saturday morning.

The problem is, the Sims 2 has turned out to be incredibly difficult to shoot with. When the Rooster Teeth gang uses Halo for machinima, the characters are mere puppets and can be posed any way the creators want. But in the Sims 2, the little virtual characters have artificial intelligence and free will. When you're playing, you do not control all the action: the fun is in putting your Sims in interesting social situations and then standing back and watching what they'll do. When Rooster Teeth's Matt Hullum builds a virtual set and puts the *Strangerhood* characters in place for a shoot, he's never quite sure what will happen. To shoot a scene in which two men wake up in bed together, Hullum had to spend hours playing with the two characters—who are nominally heterosexual—forcing them into repeated conversations until they eventually became such good friends they were willing to share a bed. Shooting machinima with Sims is thus maddeningly like using actual, human stars: they're stubborn; they stage hissy fits and stomp off to their trailers.

"We'll do three or four takes of a scene, and one of the Sims will start getting tired and want to go to sleep," Hullum said. "It's just like being on a real set. You're screaming: 'Quick, quick, get the shot! We're losing light!'"

Hullum showed me a typical *Strangerhood* scene. He put Nikki, a young ponytailed brunette in a baseball cap, in the kitchen to interact with Wade, a slacker who looked eerily like a digital Owen Wilson. (To give Wade a mellow, San Francisco vibe, Hullum programmed him to move at a pace 50 percent slower than the other characters.) Hullum pointed to Nikki's "mood" bar; it was low, which meant she was in a bad mood and wouldn't want to talk. "When they're bored, you have to lock them in a room alone for a few hours until they start to crave conversation," Hullum said. He tried anyway, prodding Wade to approach her and talk about food, one of Nikki's favorite subjects. It worked. The two became engrossed in a conversation, laughing and gesticulating wildly. "See, this footage would be great if we were shooting a scene where these guys are maybe gossiping," Hullum mused, as he zoomed the camera in to frame a close-up on Wade. Then Nikki started to yawn. "Oh, damn. See—she's getting bored. Oh, no, she's walking away," Hullum said, as the little virtual Nikki wandered out of the room. "Damn. You see what we have to deal with?"

The audience for *The Strangerhood* has not exploded the way *Red vs. Blue* did. The project is a gamble: its creators hope it will break out of machinima's geeky subculture and vault into the mainstream.

Though in a way, Hullum said, the mainstream isn't always a fun place to be, either. Before he returned to Austin to work on *Red vs. Blue,* he spent six miserable years in Hollywood working on second-rate teen movies with big budgets, like *Scooby-Doo* and *The Faculty.*

“So now to come to this, where we have total creative control of our own stuff, it’s amazing,” Hullum said, as he watched Nikki walk out of the house in search of a more interesting conversation. “I just pray we can keep this going. Because if we can’t, I’m in big trouble.”

Farhad Manjoo

Throwing Google at the Book

Google's new search engine puts a world of knowledge at our fingertips. Publishers say the Internet giant is robbing them of their rightful fees. Maybe it's time to call copyright laws history.

Just the announcement last December elicited a thrill. Google, the young oracle that brought order and sense to the World Wide Web, now planned to take on the printed word, reaching into major university libraries to scan and digitize all the knowledge contained in books. The company promised to make every printed book as accessible as a Web site, allowing anyone with Internet access to search through every page of every book for any particular word or phrase. Google had signed deals allowing it to scan millions of books at Stanford, the University of Michigan, Harvard, Oxford, and the New York Public Library. So ambitious was the effort that its only real analogues were the stuff of legend and fiction: the lost library at Alexandria and Jorge Luis Borges's fantastical Library of Babel. On hearing of Google's effort, one librarian told the *New York Times,* "Our world is about to change in a big, big way."

A year later, Google's grand plan to digitize the world's books still seems as fantastical as it did when it was first proposed. Earlier this year, the company started scanning books at libraries and on November 3 launched an elegant beta version of its book search engine—but the project faces an uncertain future.

At issue is copyright law: Does Google have the legal right to copy library books and make them searchable online? Trade groups for authors and publishers say no. In September, the Authors Guild, a professional society of more than 8,000 writers, filed suit against Google to stop the scanning project; in October, the Association of American Publishers, which represents large publishing houses, also sued. Both groups charge that Google, which does not plan to ask authors and publishers for permission before it scans their books, would engage in massive copyright infringement—and also cost the book industry a great deal of potential revenue—if it goes ahead with its effort.

Google insists that its project is legal, as it would only offer snippets—one or two sentences—of copyrighted works that publishers had not given the company permission to scan. It also argues that its plan would boost, not reduce, book sales and would be a boon to the book industry. But its quest to bring books to the Web now looks certain to spark a major courtroom battle, and it's a battle that Google, however deep its pockets and well-remunerated its lawyers, is not guaranteed to win.

"One of the great things about this conflict is it points out the absurdity of American copyright law," says Siva Vaidhyanathan, a media scholar and copyright expert at New York University. Vaidhyanathan believes that what Google wants to do may well be illegal under today's copyright regime. At the same time, he notes that Google can't really create a system that relies on publishers' granting permission to digitize their books, precisely because nobody really knows who owns the rights to all the books in the library. So, Vaidhyanathan says, Google is stuck; scanning books without asking permission may be illegal, and scanning books after asking permission is impractical to the point of impossibility.

But if copyright law stands in the way of Google's grand aim, isn't it time we thought about changing the law? That's the most salient question raised in the fight over Google's effort to build a digital library. The company—and the host of other firms that will surely follow in its path—is poised to create a tool that could truly change the way we understand, and learn about, the world around us. A loss for Google would echo throughout the tech industry, dictating not just how we use technology to improve books but also how other media—movies, music, TV shows, and even Web pages—are indexed online. Can we really afford to let content owners stand in the way of Google's revolutionary idea?

In response to the legal uncertainty, Google put its scanning project on hold for several months in the summer of 2005, but it has now resumed the project. Last week, the company put its first stash of scanned library books online. Diving into this trove is a trip. You could easily lose days in Google's digital labyrinth, not unlike the way you might walk into the stacks at Stanford or Harvard on a Friday and emerge punch drunk on a Sunday, amazed by the breadth of the work you've seen. The difference is that Google's library is searchable; you can find what you want not just by looking up an author or a Dewey decimal subject but also by typing a particular phrase or quote—"The play's the thing," say, or "Let my people go!"—that you're looking for in a book. Google will look for your search term in every page of every volume in the library and instantly show you images of the pages in each book where the phrase appears.

At the moment, Google's library mostly contains a trove of work published before 1923; copyrights on these books have expired, and the books' contents, therefore, are in the public domain, free for anyone to use in any way. Amid these titles you'll find all manner of books in Google's stash: Among many other things, there's an illustrated first edition of Henry James's *Daisy Miller,* a 1702 history of France with an exceedingly long title; *Debates of the House of Commons, 1667 to 1694,* which records a certain Mr. Finch arguing against the naturalization of aliens; and a 1785 gardening book, which advises farmers to plant hedges of holly around their corn, since "Holly does not fuck the land" and therefore rob the corn of nutrients. (It's possible, though, that this last one is actually "suck the land" and that Google's text-recognition program made a mistake with the old script.)

Google's collection also includes a vast number of books published after 1923 that publishers have already given Google permission to include—but because these books are under copyright, Google limits their functionality in order to reduce the chance that the service will negatively affect book sales. For instance, searching for the name Calliope in Jeffrey Eugenides's *Middlesex* will yield several page numbers but not the content of all those pages. That way you can't read the entire book through the search interface.

The main fight between Google and publishers involves a third category of books, those that are still under copyright but that publishers have not given Google permission to include in its library. When Google, in the course of scanning books at a library, comes upon a book published after 1923, publishers insist that the company should set it aside and get permission first; Google says that it has the right to scan these books and make them available online. The company insists that it will soon include such books in its library.

At the moment, though, what this means for you is a truncated library. Right now, no text search in Google will return any phrases contained in many popular titles. For instance, you can't find such titles as *Lolita; The Great Gatsby; The Best and the Brightest; The Da Vinci Code;* or much of anything by John Updike, Philip Roth, Richard Feynman, John McPhee, Shelby Foote, Terry McMillan, Sharon Olds, Julia Child, or Woody Allen.

This is most problematic for obscure books, books you don't know you're looking for. Take this hypothetical scenario: Let's say that somewhere in the stacks at the University of Michigan there is an essay by a writer you've never heard of, on a subject you didn't know about, in a volume no longer in print, by a publishing house no longer in business; let's say, moreover, that even though you don't really know it, this essay is exactly what you're looking for, the answer to all your searching needs, in much the same way you find Web pages every day by people you don't know that turn out to be just the thing. Ideally, as Google envisions it, you could one day go to its search engine; type in a certain bon mot; and find this book, your book. Because it's still under copyright, Google would only show you a few sentences around your search term as it appeared in the text, not the whole volume, but you'd know it was there in the library, and if you wanted it, you'd be free to check it out or find some way to buy it. Without Google's system, you'll never hear of this book.

In such a scenario, proponents of Google's plan see nothing but good—good for the company, for Internet users, and especially for authors. In most copyright disputes between content companies and tech firms, there is often a legitimate question over which party might benefit more from a new technology, notes Fred von Lohmann, an attorney at the Electronic Freedom Foundation (EFF), which sides with Google in this battle. "Take the Napster case," von Lohmann says. In that situation, Napster claimed that its file-swapping tool could increase CD sales by letting people preview music before they purchased it; the CD industry, meanwhile, said the system had caused a significant drop in sales. Both sides cited numbers to support their arguments, and each theory sounded at least plausible.

"But with the Google Print situation, it's a completely one-sided debate," von Lohmann says. "Google is right, and the publishers have no argument. What's their argument that this harms the value of their books? They don't have one. Google helps you find books, and if you want to read it, you have to buy the book. How can that hurt them?"

Obscure books—books that are out of print or otherwise hard to get hold of—would stand to gain the most from such a system, and it turns out that there are plenty of such books in the libraries Google plans to scan. Not long ago, the Online Computer Library Center, a nonprofit library research group, set out to count and catalog the books Google would capture in its project. The OCLC determined that at the five research libraries with which Google had formed deals, about 80 percent of the books in the stacks were published after 1923 and still under copyright. But only a small number of these books are currently in print.

Tim O'Reilly, a computer book publisher and sponsor of influential tech conferences, points out that in 2004 only 1.2 million different book titles were sold in the United States, according to Nielsen Bookscan. This means that while a significant number of library books are protected by copyright, they are also out of print—70 percent or more, O'Reilly estimates. These books, he says, represent the "twilight zone" of the publishing world; someone owns them, but since they're perceived to have no commercial value (because they're no longer sold in stores), publishers don't have any incentive to promote and market them, let alone to go through the expense of scanning them and making them searchable online.

Indeed, in many cases the publishers and rights holders of these books are unknown. There is no national registry of copyright holders in the United States, as there is a national registry of patents. Any book published is automatically granted a copyright, and if a book publisher goes out of business, or an author dies, the copyright to the work may well be buried in contracts that long ago turned to dust. "We precluded any possibility of creating a copyright database," says Vaidhyanathan, and "it's impossible for a company like Google, or a historian, or a documentary filmmaker, or anyone to find out who owns what. Even publishers don't know what they own. It's just impossible."

O'Reilly is one of few publishers who support Google's plan, and he likes it precisely because he thinks it will shed light on these little-known titles whose rights holders are hard to track down. "One of the biggest arguments for Google's approach is that it is the only solution that solves a hard problem," O'Reilly says. He points out that only 2 percent of books sold in 2004 had more than 5,000 copies purchased; the rest languished in obscurity. And that, he wrote in a recent *New York Times* Op-Ed, "is a far greater threat to authors than copyright infringement, or even outright piracy." Google, O'Reilly went on to write, "promises an alternative to the obscurity imposed on most books. It makes that great corpus of less-than-bestsellers accessible to all. By pointing to a huge body of print works online, Google will offer a way to promote books that publishers have thrown away, creating an opportunity for readers to track them down and buy them . . . In one bold stroke, Google will give new value to millions of orphaned works."

But if it's true that Google's new system would be good for old books, it's also true that the system would be a good one for Google, helping to cement its position as the world's dominant search engine. Nobody knows—and Google isn't saying—how much money the company stands to make directly from the library venture. In a recent Op-Ed in the *Wall Street Journal,* Eric Schmidt, Google's CEO, sought to play down the company's profit motive. He pointed out that the company will not place advertisements on search pages for books it scans from libraries. Though Google will place ads on pages for books that publishers have given the company permission to include, it will send publishers the "majority" of revenue for such ads, Schmidt wrote. Google will include a referral link to let people purchase books they find in the library—a "Buy this book" link to several major online bookstores—but the company won't "make a penny on referrals," he wrote.

Rather than making money from the individual book searches, Google's library will pad Google's bottom line by increasing the value of its main search engine. Although Google remains by far the world's most popular search engine, it faces stiff competition from other firms—Yahoo, Microsoft, Amazon, and others—who want a share of its vast audience and are also planning ventures to digitize and offer search systems for print books and other media. By offering something—millions of books—that others are not yet offering, Google will be creating another reason for users to stick with its interface for searching the Web.

But Google's competitors are not far behind. Amazon, which already offers a feature to search inside many books in its store, has just announced a plan to let users buy specific pages of books. Microsoft and Yahoo, meanwhile, have joined the Open Content Alliance, a nonprofit group that includes contributions from the Internet Archive and the University of California at Berkeley and that plans to digitize books only after asking for publishers' permission.

It's Google's profit motive that raises the suspicion of authors and publishers. As they see it, digital technology provides authors and publishers a new way to make a great deal of money on their back catalogs of books—a huge source of revenue that is currently being untapped. Google is creating a system that exploits that back catalog, so why shouldn't Google pay content owners for the right use of that catalog?

"The author is creating the value here," says Paul Aiken, executive director of the Authors Guild, "and the author should get some of the money. If there's a new value for books created on the Internet, the authors should be given new incentives to create works for it."

Aiken compares Google's plan to use books with the way Hollywood uses novels as plots for its movies. When film producers first started making movies from books, "They could have said, 'Hey, how does it hurt the author if I make a movie from his book?'" Aiken points out. "You could argue, after all, that more people would buy the book because of the movie." But that's not the way the world works, Aiken says. Hollywood pays publishers for the rights to novels they want to use, and in the same way, Google should pay publishers—who would then distribute money to authors—for the right to add books to its database. Aiken declined to offer a detailed, specific plan by which Google could pay authors for their contributions to its search engine. But he suggested that one idea might look very much like the system that radio stations use to pay musicians. Google could pay an annual licensing fee to publishers, and the money would be distributed to publishers and authors according to how often those books were viewed in the search engine.

Aiken's argument is echoed by publishers. Google, notes Pat Schroeder—the former Democratic congresswoman from Colorado who now heads the Association of American Publishers (AAP)—is rich! Both Schroeder and Aiken fingered Google's latest earnings report, which showed that the company recorded a 700 percent increase in profits in the third quarter of 2005 compared with the same quarter in 2004. In 2004, the company's revenues exceeded $3 billion and the firm could make double that much by the end of the current fiscal year. In other words, Google is sitting on a gold mine, and authors and publishers, notably, are not. "They try to sound like they have this high moral purpose so they can't be bothered with permission," Schroeder says of Google. "They tell us it's good for the world, and it's good for publishers. The thing they leave out is it's really good for Google."

The Authors Guild hasn't conducted a survey of its members to determine what they think of the Google plan, but Aiken says the e-mails he gets from authors run overwhelmingly in support of the guild's lawsuit against the company. There's no reason not to believe Aiken; it's not hard to find authors who are deeply suspicious of what Google plans to do. Take Peter Salus, a veteran author of computer books who lives in Toronto. Salus has authored, coauthored, or edited about two dozen books, some of which are in print and some of which are not. He says that he understands the benefits of Google Print—but he just wants the company to do one thing in return: ask his permission.

"I think it's absurd that they think the authors should have to come to them to opt out of the database," rather than the other way around, Salus says. Because Google's project directly benefits from his and his fellow authors' work, Salus says, it's incumbent upon the company to make sure that authors are OK with what it's doing. And what if it's too logistically difficult for Google to find every author of every book in the library and ask his or her permission? "That's tough—it really breaks my heart," he says. "But there is no burden on me or anybody else to make it easier for them to make money."

Many authors feel differently. One is Julian Dibbell, author of *My Tiny Life,* a memoir of the author's life in the virtual computer world called LambdaMOO. When told of Aiken's theory that Google's database would use authors like him in the same way that Hollywood might use them and that authors should get paid for allowing their books to go to Google, Dibbell said, "My blood is boiling just as you relay this to me." As Dibbell sees it, "Google is not piggybacking on my creative effort in the same parasitic way that a movie based on a novel might be doing." To Dibbell, Google is acting not like the Hollywood producer who steals an author's ideas but instead like a book reviewer who popularizes an author's work. After all, Dibbell notes, book reviewers routinely use snippets from books in their reviews, and magazines and newspapers make loads of money from advertisements they run alongside book reviews. Authors don't feel entitled to any of that money, he says, so why should they get a slice of the money Google will make from its service? "Given what's at stake here, which is the creation of a resource that nobody is denying is a good thing, their stance seems wrong to me," Dibbell says of his fellow authors.

Whether Google is acting more like a book reviewer or like a movie producer in its use of other people's books may turn out to be a key question in the legal battle to come. Google, which did not make a company attorney available to *Salon,* has insisted that, as with book reviewers, copying books falls within the "fair use" exception of American copyright law. Google essentially argues that because it is copying books as a step toward a larger goal—the creation of a search engine of library books—its actions are permitted. After all, the company points out, it does exactly the same thing with Web pages. To create a searchable index, the Google Web search engine copies entire Web sites—all of *Salon,* for instance, resides in Google's servers—without their permission. If copying a Web site is OK, why is copying a book not?

In his *Wall Street Journal* editorial, Google CEO Schmidt defended the company's legal interpretation. "The aim of the Copyright Act is to protect and enhance the value of creative works in order to encourage more of them—in this case, to ensure that authors write and publishers publish," he wrote. "We find it difficult to believe that authors will stop writing books because Google Print makes them easier to find, or that publishers will stop selling books because Google Print might increase their sales."

Fred von Lohmann of the EFF agrees with Google's view of the law, and he says that several federal court rulings uphold what Google is doing. Courts have already ruled, for instance, that it's OK for companies to make copies of video games as part of their efforts to reverse-engineer those video games; because the copied video games weren't meant to be sold and were instead used for some other purpose (the creation of a reverse-engineered product), the copies were considered fair use. Then there's *Kelly v. Arriba Soft,* a 2002 case in which a judge ruled that it was legal for a search engine company to copy photographs from other sites online and display "thumbnails" of those photos as part of its search results. The thumbnails, the court said, were quite different from the original photographs—they were smaller and of lower resolution—and were unlikely to be used for the same purposes as the originals.

A similar argument can be made on behalf of Google's book search engine, von Lohmann points out. Google is not giving readers the exact copy of the books it scans from the library; rather, it's just giving them a snippet in the same way that a graphical search engine may show users a thumbnail image of a picture.

But if there are cases that support Google's view of copyright law, there are also federal court cases that line up against it. The main case involves MP3.com, a boom-time Internet company that copied tens of thousands of CDs to its internal database without getting record labels' permission to do so. (MP3.com planned to make digital tracks of songs available to anyone who could prove they'd purchased a physical CD of the music.) In 2000, a federal judge in New York ruled that MP3.com's copying of CDs without permission infringed upon the record labels' copyrights. A reasonable judge may look at Google's actions as being essentially no different; just as MP3.com copied CDs without asking, so too is Google copying books.

Both authors and publishers sued Google in the Southern District of New York, where the MP3 case is still an important legal precedent; the choice of locale, says Vaidhyanathan, was not an accident. "They picked the court that is likely to rule along the lines of the MP3.com case and less likely to think that *Kelly v. Arriba Soft* was a good decision," he says.

Aiken of the Authors Guild, for one, is sanguine about his lawsuit's prospects. "I don't think if you were to survey copyright lawyers you'd find their view prevailing," he says of Google. "Our case is very, very strong. It would shock the copyright bar if it was decided against us." Still, both Aiken and Schroeder say they are open to settlement discussions with Google. "There's a lawsuit," says Aiken, "but if they wanted to negotiate a license we could work something out."

Google has not yet filed a legal response in either case, and it's unclear whether the company is open to a settlement. But several observers say Google may also be inclined to think about sitting down with the authors and publishers and talking about the case, including discussing a possible permission-based system for getting books into its library. For one thing, Google wouldn't want to risk losing in this case, says Vaidhyanathan, as a badly worded ruling in this case could put its other operations in legal jeopardy.

O'Reilly says that he was recently at an event sitting between Larry Page, one of Google's cofounders, and John Sargent, who sits on the board of the AAP. "I'm not going to tell you what they said," O'Reilly says, "but I think it's fair to characterize what they're doing now as negotiating by lawsuit and press release." Each side is hoping for some early legal decisions to go its way, O'Reilly says, and then the real settlement talks will begin.

But whatever happens with Google's venture, a more lasting outcome from this case may be a change in the way we think about how much control an author or publisher, musician or record label, filmmaker or studio is allowed to exert over works they create—a question that has been cast into stark relief in the digital age.

Lawrence Lessig, a Stanford law professor and copyright scholar, likes to tell the story of Thomas Lee and Tinie Causby, two North Carolina farmers who in 1945 cast themselves at the center of a case that would redefine how society thought of physical property rights. The immediate cause of the Causbys' discomfort was the airplane; military aircraft would fly low over their land, terrifying their chickens, who flew to their deaths into the walls of the barn. As the Causbys saw it, the military aircraft were trespassing on their land. They claimed that American law held that property rights reached "an indefinite extent, upwards"; that is, they owned the land from the ground to the heavens. If the government wanted to fly planes over the Causbys' land, it needed the Causbys' permission, they insisted.

The case, in time, came to the Supreme Court, where Justice William O. Douglas, writing for the Court, was not kind to the Causbys' ancient interpretation of the law. Their doctrine, he said, "has no place in the modern world. The air is a public highway, as Congress has declared. Were that not true, every transcontinental flight would subject the operator to countless trespass suits. Common sense revolts at the idea. To recognize such private claims to the airspace would clog these highways, seriously interfere with their control and development in the public interest, and transfer into private ownership that to which only the public has a just claim."

Google supporters say the publishers' objection resembles that of the Causbys'. Just as the airplane rendered the Causbys' rights to the skies incompatible with the modern world, the Web has rendered publishers' right to the digital universe out of tune with modern technology and society. The public benefit of making millions of books, or excerpts of books, readily available to people worldwide "could be the most important contribution to the spread of knowledge since Jefferson dreamed of national libraries," Lessig wrote recently on his blog. "It is an astonishing opportunity to revive our cultural past, and make it accessible. Sure, Google will profit from it. Good for them. But if the law requires Google (or anyone else) to ask permission before they make knowledge available like this, then Google Print can't exist." And if Google Print can't exist, maybe it's time to reexamine the law.

Steven Johnson

Why the Web Is Like a Rain Forest

Software upgrades promise to turn the Internet into a lush rain forest of information teeming with new life.

Some technological revolutions arrive as revelation. You hear a human voice wafting out from a rotating plastic disk or see a moving train projected onto a screen, and you sense instantly that the world has changed. For many of us, our first encounter with the World Wide Web a decade ago was one of those transformative experiences: You clicked on a word on the screen, and instantly you were transported to some other page that was served up from a computer located somewhere else, across the planet perhaps. After you followed that first hyperlink, you knew the universe of information would never be the same.

Other revolutions creep up with more subtlety, built of tweaks and minor advances, not radical breakthroughs. E-mail took decades to gestate, but now many of us can't imagine life without it. There's a comparable quiet revolution under way right now, one that is likely to fundamentally transform the way we use the Web in the coming years. The changes are technical and involve thousands of individual programmers, dozens of start-ups, and a few of the largest software companies in the world. The result is the equivalent of a massive software upgrade for the entire Web, what some commentators have taken to calling Web 2.0. Essentially, the Web is shifting from an international library of interlinked pages to an information ecosystem, where data circulate like nutrients in a rain forest.

Part of the beauty and power of the original Web lay in its simplicity: Web sites were made up of pages, each of which could contain text and images. Those pages were able to connect to other information on the Web through links. If you were maintaining a Web site about poodles and stumbled across a promising breeder's home page, you could link to the information on that page by inserting a few simple lines of code. From that point on, your site was connected to that other page, and subsequent visitors to your site could follow that connection with a single mouse click. In some basic sense, those two pages of data were interacting with each other, but the exchange between them was rudimentary.

Now consider how a group of poodle experts might use the Web 2.0. One of them subscribes to a virtual clipping service offered by Google News; she instructs the service to scan thousands of news outlets for any articles that mention the word *poodle* and to send her an e-mail alert when one of them comes down the wire. One morning, she finds a link to a review of a new book about miniature poodles in her inbox. She follows the link to the original article, and using a standard blogging tool like TypePad or Blogger, she posts a quick summary of the review and links to the Amazon page for the book from her blog.

Within a few hours of her publishing the note about the new book, a service called Technorati scans her Web site and notices that she has added a link to a book listed on Amazon. You can think of Technorati as the Google of the blog world, constantly analyzing the latest blog posts for interesting new developments. One of the features it offers is a frequently updated list of the most talked-about books in the blog world. If Technorati stumbles across another handful of links to that same poodle book within a few hours, the poodle book itself might show up on the hot books list.

After our poodle expert posts her blog entry, she takes another few seconds to categorize it, using an ingenious service called del.icio.us, which tags it with her content-specific title, like "miniature poodles" or "dog breeding." She does this for her own personal use—del.icio.us lets her see at a glance all the pages she has classified with a specific tag. But the service also has a broader social function: Tags are visible to other users as well. Our poodle expert can see all the pages that other users have associated with dog breeding. It's a little like creating a manila folder for a particular topic, and every time you pick it up, you find new articles supplied by strangers all across the Web.

Del.icio.us's creators call the program a social bookmarking service, and one of its key functions is to connect people as readily as it connects data. When our poodle lover checks in on the dog-breeding tag, she notices that another del.icio.us user has been adding interesting links to the category over the past few months. She drops him an e-mail and invites him to join a small community of poodle lovers using Yahoo's My Web service. From that point on, anytime she discovers a new poodle-related page, he'll immediately receive a notification about it, along with the rest of her poodle community, either via e-mail or instant message.

Now stop and think about how different this chain of events is from the traditional Web mode of following simple links between static pages. One small piece of new information—a review of a book about poodles—flows through an entire system of reuse and appropriation within hours. The initial information value of the review remains: It's an assessment of a new book, no different from the reviews that appear in traditional publications. But as it ventures through the food chain of the new Web, it takes on new forms of value: One service uses it to help evaluate the books with the most buzz; another uses it to build a classification schema for the entire Web; another uses it as a way of forming new communities of like-minded people. Some of this information exchange happens on traditional Web pages, but it also leaks out into other applications: e-mail clients, instant-messenger programs.

The difference between this Web 2.0 model and the previous one is directly equivalent to the difference between a rain forest and a desert. One of the primary reasons we value tropical rain forests is because they waste so little of the energy supplied by the sun while running massive nutrient cycles. Most of the solar energy that saturates desert environments gets lost, assimilated by the few plants that can survive in such a hostile climate. Those plants pass on enough energy to sustain a limited number of insects, which in turn supply food for the occasional reptile or bird, all of which ultimately feed the bacteria. But most of the energy is lost.

A rain forest, on the other hand, is such an efficient system for using energy because there are so many organisms exploiting every tiny niche of the nutrient cycle. We value the diversity of the ecosystem not just as a quaint case of biological multiculturalism but because the system itself does a brilliant job of capturing the energy that flows through it. Efficiency is one of the reasons that clearing rain forests is shortsighted: The nutrient cycles in rain forest ecosystems are so tight that the soil is usually very poor for farming. All the available energy has been captured on the way down to the earth.

Think of information as the energy of the Web's ecosystem. Those Web 1.0 pages with their crude hyperlinks are like the sun's rays falling on a desert. A few stragglers are lucky enough to stumble across them, and thus some of that information might get reused if one then decides to e-mail the URL to a friend or to quote from it on another page. But most of the information goes to waste. In the Web 2.0 model, we have thousands of services scrutinizing each new piece of information online, grabbing interesting bits, remixing them in new ways, and passing them along to other services. Each new addition to the mix can be exploited in countless new ways, both by human bloggers and by the software programs that track changes in the overall state of the Web. Information in this new model is analyzed, repackaged, digested, and passed on down to the next link in the chain. It flows.

This is good news whether we love poodles or not, but it's also good news economically because the diversity of the ecosystem makes it a fertile environment for small players. You don't have to dominate the food chain to get by in the Web world; you can find a productive niche and thrive, partially because you're building on the information value created by the rest of the Web. Technorati and del.icio.us both began as small projects created by single programmers. They don't need huge staffs because they're capturing the information supplied by the countless number of surfers who use their services, and they're building on other tools created by other people, whether they work in a home office or in a vast international corporation like Google. All of which makes this the most exciting time to be on the Web since the glory days in the mid-1990s. And the revelations aren't about to stop. As we figure out new ways to expand the complex information food chains of Web 2.0, we will see even more innovation in the coming years. Welcome to the jungle.

Lisa Margonelli

China's Next Cultural Revolution

The People's Republic is on the fast track to become the car capital of the world. And the first alt-fuel superpower.

The Challenge Bibendum is the anti-Nascar, a road rally where dozens of cars, two-wheelers, and buses vroom the straightaways like a pack of DustBusters, cough out water vapor instead of sooty exhaust, and corner at peak speeds of 35 mph. Named for the morbidly obese mascot of Michelin, which sponsors the event, Bibendum is the proving ground for alternative-fuel and low-emissions vehicles.

For the first five years of its existence, the rally was staged in rich cities with bohemian tendencies—San Francisco, Heidelberg, Paris. But last fall Michelin brought the Bibendum to Shanghai. The booming Chinese auto market, which grew by 76 percent in 2003, is an obvious lure. It's a market still under central control—for the moment, anyway—which means that if Beijing wants to go green, it can go in a huge way. And so in Shanghai, Bibendum lost its utopian vibe. The stakes were simply too big: What will 1.3 billion people drive?

The answer, believes professor Huang Miao Hua, is an electric car prototype made by her students at the Wuhan

University of Technology. The Aspire (not to be confused with the Ford compact car) is a giddy marriage of tadpole and pickup truck. The $12,000 target price includes a Linux OS, GPS, and an onboard bicycle. A bike? If you get stuck in gridlock, Huang explains, you can park the car and pedal instead. Think of it as a concept car for traffic jams. She pushes up the Aspire's door (it opens vertically, for parking in tight spots) and smiles. "Get in," she says. As the vehicle lumbers to a start, engine whining under the strain, the driver shouts, "It's got a few problems, but it feels good, doesn't it?"

In the West, clean cars mostly have been the toys of wealthy worrywarts—too expensive to be economical and too technically challenged to be cool. China's feeling an urgency that slower-growing countries don't face. The demand for oil is skyrocketing, rising even faster than the price. And here's the eye-opening stat: In the absence of new regulations, pollution-related illness will suck up as much as 15 percent of the country's gross domestic product by 2030.

China's central planners are throwing everything at the problems of fuel and pollution—hybrids, electric cars, propane taxis—all while building conventional cars and infrastructure at a furious pace. "There's a controversy about 'Green GDP' and how to grow," says He Dongquan, a transportation expert at the Energy Foundation in Beijing. "China's in a transition where everyone's mind is changing." Amid the hurly-burly, the only thing that's clear is the future, where hydrogen beckons.

China is already taking bold steps toward an alt-fuel future. In late 2003, Beijing mandated some of the world's toughest fuel-efficiency standards. China is even now one of the largest markets for alternative fuel vehicles, with 200,000 in service. In preparation for the 2008 Olympics, Beijing officials plan to convert their entire bus fleet of nearly 120,000 vehicles to run on compressed natural gas (CNG).

All this opens up vast opportunities for automakers. The major car manufacturers (with the exception of Honda) have come to Bibendum to show that they're ready to play China's game, whatever it turns out to be. Toyota will begin producing hybrid Priuses in Changchun by the end of the year. GM, which made 15 times more profit per sale in Asia than at home in 2003, will manufacture hybrid buses for Shanghai. "This will be the biggest market in the world by 2010," says Dongfeng Citroen chief Gilles Debonnet, standing beside a CNG car his company designed for Bibendum. "If we don't bring a [low-emissions] solution to the taxi market, then we can't stay."

Decades behind developed nations when it comes to supporting a car culture, China may actually benefit from its very backwardness. All those bicycles mean there isn't a cumbersome—and entrenched—gasoline infrastructure to stand in the way of the next big thing. That's why China hopes to eventually bypass the oil-based auto culture and go right to a hydrogen economy. "Some theorists believe China has an advantage with fuel cells because it has no resistance," says General Motors vice president David Chen as he attends to a Shanghai dignitary at Bibendum. "It's been cut off from the world for 30 years. It may be in a unique situation to leapfrog."

Leapfrogs are an intoxicating vision, but can this one really jump? "We consider China a wild card," says Shell Hydrogen VP Gabriel de Scheemaker, who installed Iceland's hydrogen infrastructure and is now at Bibendum trying to get into the Chinese market. His eyes get dreamy as he imagines Shanghai on H_2—city blocks powered by fuel cells, cars filled from hydrogen supplies embedded in buildings: "In Deng [Xiaoping]'s day, he experimented with whole cities!"

Although China's in an experimental mood, innovations are hard to finance. The Aspire bobbles toward Shanghai's Formula 1 track, past Toyota's Prius; Volvo's lozenge-shaped 3CC concept car; and the Mercedes-Benz A-Class F-Cell, which has its magnificent fuel cell guts jammed into a frumpy hatchback. The team from Wuhan is taking on the big guys with two goofy Aspires made for a total of $60,000.

The six-day Bibendum turns out to be a coming-out party for China's homegrown clean cars. The Aspire wins a special design prize. And of the 43 Chinese vehicles entered, 19—including 11 two-wheelers, 6 buses, and 2 cars from an array of fuel sources—score very high on the Bibendum tests of overall emissions, CO_2 emissions, noise, fuel economy, braking, slalom, and acceleration. A year before, there wasn't even one Chinese entry. "We wanted to do something good for the country," Huang says, her students giggling with excitement as they push the little car out of the garage for the trip back to Wuhan. "My students gave without expecting any return. That's the spirit!"

Here's the new cultural revolution: Every morning Wang Jian Shuo and his wife leave their condo in the suburbs of Shanghai, get into their Fiat sedan, and drive to jobs in the city. Two years ago, they lived in a cramped, decrepit apartment in the center of Shanghai, and Wang, an engineer for Microsoft, traveled to work by bus or train. "I never thought of getting a car," he says. "Driving was a very serious profession—like medicine." Cars were for party bureaucrats or at least the very rich.

But in 2000, Shanghai's per capita GDP (already much higher than China's overall) rose above $4,000, and the roads started filling with private cars. Local highways, which were designed by engineers who'd never driven, clogged. Shanghai's narrow streets became so congested that commuters abandoned their bicycles for the subway just to avoid the cars. Smog grew so thick that on many days you couldn't even see the boisterous skyscrapers looming above you.

And so, a year ago, Wang moved into a spacious condo in the suburbs—and bought a car. "The change the car brings my life is bigger than the house," he says. "My life scope is much larger now." Today Wang and his wife shop in Western-style supermarkets instead of haggling with the fishmonger, and they can drive to visit friends and return home by car long after the subway has shut down for the night. They grew up in a world bounded by transit schedules, shabby housing, and nosy neighbors, but now they live in an airy apartment, surrounded by the brand-new highrises that have sprung out of the rice paddies. Some nights, when they're tired, Wang and his wife get in the car and drive out to the new airport just to experience speeding down the empty highway. But even that road is filling up. It makes Wang happy he bought a car as soon as he did. "When a car becomes something everyone can afford, forget it," he says. "You won't be able to drive."

At a Hyundai dealership not far from Wang's condo, families prowl the showroom, inspecting the stitching on the seats, criticizing the design of the rear lights, trying to find the biggest car for their yuan. A TV blares a government program featuring a singer in a yellow dress crooning in front of a suburban development. "Nowadays life is getting better, sweeter and sweeter," she sings. "You can fulfill your dreams. The roads are getting wider and wider."

Managing dreams is a big problem for the Beijing bureaucrats who pull the levers of China's economy. Yang Yiyong is low enough in the party hierarchy that he'll talk with a foreign reporter, high enough that he insists on meeting in the back room of a restaurant famous for its duck with stewed fruit. His official title is deputy director of the Institute of Economic Research, which is a government-sponsored think tank.

Yang wears a serious pin-striped suit and talks big numbers. China's population, he says, will approach 1.5 billion in 2030. The only way to forestall economic calamity is to maintain 25 consecutive years of high annual GDP growth. That kind of growth, in turn, requires massive amounts of energy. Already the world's second-largest oil importer, China is expected to more than double imports by 2020. This is a painful subject for Yang, who fulminates against cars, car culture, traffic, and the prestige New China is attaching to big cars. "I object to this vague notion of status," he says.

His concern is ideological, but the problem is practical. After food, oil is the most important issue for Chinese economic planners. Without an increasing supply of oil, high GDP growth will be impossible, creating unemployment and social unrest, potentially threatening the government's hold on power. That's not all. Dependency on foreign oil, in Yang's opinion, inevitably leads to war. Every official I interview makes the same point. Yang uses a pun to summarize the leadership's view: "If you pump for oil, you have to fight wars for it." (Pump and fight sound similar in Mandarin.)

In the face of an oil crisis, the government is embracing fuel efficiency and alternative energy resources. In every scenario, oil imports will rise, but the hope is that new technologies and conservation will minimize the rate of growth. The plan is to replace 10 percent of China's energy supply with renewable sources by 2010, 12 percent by 2020. (Today, less than 1 percent comes from renewables.) "We're not saying we can reduce consumption," he cautions, "but we can reduce the increase and win some time."

A chauffeur-driven Audi A6 stops near the Mao statue facing the gates of Shanghai's Tongji University, and Wan Gang steps lightly out of the back. He is a compact man in his early 50s who retains the enthusiasm and pink cheeks of a boy genius. As the chief scientist of the 863 Program's Key Electric Vehicle Project (the 863, named for its approval date, March 1986, is China's national high tech R&D initiative), Wan has to get Chinese industry mass-producing fuel cells by 2020. It's an ambitious national agenda that started in 2001 with an unambitious budget: $106 million. That figure must sustain the network of 200 universities and companies that are developing and testing scores of electric, hybrid, and fuel cell vehicles.

The fuel cell mission is born partly out of necessity. In 2000, China's Ministry of Science and Technology contacted Wan, who had been living in Germany for a decade doing research for Audi. The ministry asked him to come back and create a strategy for the overall Chinese auto industry. Wan concluded that it would be futile to try to compete with the West by building a better or cheaper internal combustion engine. Getting a head start in fuel cell technology would be the country's best bet. But still a long bet.

Wan, who is also the president of Tongji University, convenes his interview with me in a giant Mao-modern formal room. The tractor-sized chairs inhibit normal conversation, so he quickly moves us to a utilitarian conference room, indistinguishable from one you might find in an office in Berlin or New York. Wan lays out a 15-year plan that will lead to fuel cell cars, putting China at the forefront of the hydrogen economy. He pulls out a piece of paper. "I'm trying to demonstrate that the picture is reasonable and practical," he says, sketching a grid.

The grid contains four major fuel types: electric, hybrid, CNG, and hydrogen. Hydrogen, Wan explains, is a glorified battery, a way to store energy from various sources—coal, solar, nuclear, or hydroelectric—until it's needed. He draws a circle lassoing the hydrogen and electric columns. Today's investments in electric car R&D, he argues, will still be paying off in a hydrogen fuel cell–powered future.

Likewise, hybrid technology, Wan explains, is all about fuel efficiency. For example, advanced hybrids brake by forcing the electric motor to spin backward, generating energy that's stored in the battery. "Engineers in the States say hybrids are transitional, but I believe the technology will last a long time," he says, drawing arrows across the grid to show how regenerative braking technology will make both electric- and hydrogen-powered cars more efficient.

CNG cars will require a network of gas pipes connecting refineries to filling stations. But natural gas, he explains, can be converted easily to hydrogen. And with one final pencil stroke, the whole chicken-and-egg problem of hydrogen cars versus hydrogen infrastructure is gone. Wan holds up the grid, covered in optimistic arrows, and declares, "China has the advantage of not being burdened by previous investment."

China is waving its big red wand, but will a hydrogen economy pop out of this hat? "It's lovely to forecast out 15 years," says John Wallace, an American fuel cell consultant working with clients in China, "but nobody remembers." Wallace is fond of Wan Gang and admires the 863 Program's "credible" technology and pluck. But he says no amount of determination can summon the resources that China needs to make hydrogen vehicles a reality: start-up infrastructure, niche technology companies, and venture capital firms.

Those resources may be coming. Venture capitalist Mike Brown, chair of Canadian fuel cell investment firm Chrysalix Energy, is looking at China. "Wan's plan is eminently doable. If they went balls to the wall, they could do even more," Brown says. "The big question is whether the government will have the nerve to scoop the world."

In the lobby of one of Shanghai's vast Epcot Center–like hotels, Cai Xiaoqing taps his foot restlessly. He wants to jump-start the hydrogen economy immediately. With an astronaut's brush cut of salt-and-pepper hair, Cai looks the part of the former space program technocrat he is. As director of the Equipment Industry Department for Shanghai's Municipal Economic Commission, his job is to make Shanghai the Detroit of China.

Like everyone else here, Cai speaks in billions and of far-off years, but he's more impatient than most. He can't wait for a homegrown fuel cell. Cai wants Shanghai to quickly move to hydrogen. But how do you start a hydrogen economy without a hydrogen car?

Cai looks abroad and sees foreign auto manufacturers sitting on piles of expensive fuel cell technology with nowhere to test it. In California, they've been reduced to clownish stunts like putting a fuel cell in Arnold's Hummer. Cai can do better than that.

Bouncing slightly, Cai pitches Shanghai as a test track: 10 fuel cell cars in circulation by the end of this year, 1,000 by 2010, and 10,000 by 2015. But making hydrogen cars a reality by 2020 will require government investment in technology and subsidies to consumers. Cai calls it "a long step." Others say it's impossible. But consider the payoff: clean cars ready for export just as the rest of the world starts to choke on pollution and gasoline supply problems.

To provide the fuel cells, Cai has his eye on General Motors, which has poured more than a billion dollars into a hydrogen-powered fleet but has nowhere to drive it. "If China develops the infrastructure, GM would put those cars to use," Cai says. "I think they see China's big market, too."

In fact, they do. For more than a year, Tim Vail, GM's director of business development in charge of commercializing fuel cells, has been traveling to China and liking what he finds. He looks at Shanghai's propane taxis, 38,000 in all, and sees an industry ready to experiment. He looks at Shanghai's $1 billion magnetic levitation or "maglev" train and sees a city that's ready to spend. He looks at a coal-processing plant in the city and sees a source of industrial hydrogen that should last for the next 15 years. But most important, he sees a government that's ready to do the social engineering needed to speed the adoption of fuel cells. To Vail, Shanghai's ridiculously crowded city center, where the nouveau riche compete to conspicuously outconsume each other, is a plus. "You would see well-heeled people buying fuel cell cars if they had enhanced rights," he says. "More than anywhere else, Shanghai could say, 'Only fuel cell vehicles [in the downtown]' without a lot of debate."

Last October, GM chair Rick Wagoner shook hands with the vice mayor of Shanghai. They agreed to codevelop a fuel cell demonstration vehicle and help write the standards and policies for hydrogen power and infrastructure. Meanwhile, Volkswagen endowed a chair at Shanghai's Tongji University and agreed to jointly research fuel cell technology. "It's strategic positioning at this point," says Chris Raczkowski, a top Beijing-based alternative energy consultant, "but some companies may get a captive market for their products, and that's really the only way to get a market jump-started."

The day after the GM deal is struck, local dignitaries gather at the Shanghai International Automobile City in Jiading to celebrate this triumph of focus and vision. Four years ago, Jiading was a suburban farming village. Out went the farmers; in came the $300 million F1 racetrack (site of the Challenge Bibendum), Tongji University's College of Automotive Studies, six square miles of automotive-themed industrial park, and a golf course.

A band plays "Remember the Red River Valley," and Wan Gang takes the stage to reminisce about the eight years he spent in the countryside during the Cultural Revolution. Back then his work crew built an entire town from scratch: the roads, the electrical grid, farms, even a hospital. Yesterday they built Motor City. Tomorrow they'll build a hydrogen economy.

Across the hall, the 863 Program unveils its newest prototype, the Spring Light 3, a fuel cell–electric hybrid with steer-by-wire technology and regenerative braking. Target price: about $5,000—the car for the new masses. While Western automakers often boast that their enviro-wagons make "no compromises," the 863 Program makes compromise its guiding principle. Like the funky Aspire, the Spring Light takes you where you want to go, without promising more. American cars are all ego, but the Aspire and Spring Light are friendly, even neighborly. They're all about getting along, not getting away.

By the end of the afternoon at Jiading, it isn't the Spring Light or the VIPs that are making the big impression. It's Wan's preview of Tongji's new dormitories, complete with hot water and Western-style toilets. The engineering students see the bathrooms and let out a loud gasp. Their reaction is part awe, part appreciation, part anticipation of a new world that can only be better. Does the hydrogen highway start here? Maybe. Maybe your future and mine is being created by people desperate enough to imagine it.

Mike Daisey

The Coil and I

Adventures with a mad scientist's lightning machine

I require almost nothing for my monologues: a table, a glass of water, a chair. But Alex, the artistic director of Les Freres Corbusier—a downtown theater troupe that recently put on a production in which Stalin, Goebbels, and FDR were represented by live rabbits—insists that I ask for whatever I need to realize my vision. What could I possibly need? Dancers? More rabbits? "Maybe a Tesla coil?" I asked.

This is how Alex and I find ourselves standing in a small, dark room in Tribeca next to a very large Tesla coil. It's the largest in Manhattan and, if some Internet discussion boards are to be believed, the largest on the eastern seaboard. It is taller than I am, with a broad, boxy base studded with large, dangerous-looking white insulators. Growing out of the base is a tall, copper-red cylinder of wire coil, wrapped by hand thousands of times to create an enormous electromagnet. On top of the cylinder sits a bright, bare, metal toroid, the discharge point. The Tesla coil looks exactly like what it is: a lightning-throwing death machine.

My latest monologue is about the war over electricity standards between Nikola Tesla and Thomas Edison.

Tesla, the coil's inventor, was a madman and genius whose discoveries underpin much of our current understanding of electromagnetism. Sadly, his life slid downhill until he died penniless and delusional, filling his days writing sonnets to the pigeons on the roof of the Hotel New Yorker. On his descent to madness, Tesla embodied the archetype of the mad scientist. In fact, in the classic Frankenstein movies, the machine that brings the monster to life is a modified Tesla coil known as a Jacob's ladder.

When active, the Tesla coil alternates electrical current far more than the 60 times per second in standard AC. The coil instead flips the current faster and faster until it is alternating millions of cycles per second. When that happens, lightning will pour out of the toroid and fill the room.

Tesla coils and other high-voltage electrical generators look spectacular but have no conventional use. At the same time, Tesla's induction motor—the core of the Tesla coil—is used in pretty much every kitchen appliance and household device, and the principles at work in the Tesla coil are used to create the high voltages needed for conventional television picture tubes to work. Still, only two groups are doing much work with extremely high–voltage electrical effects these days. First, there is the military, which has experimented for years with using ultra-high-amplitude electricity as a weapon. Second are hobbyists, amateur physicists, and adventurous people who want to share in what Tesla must have felt: harnessing lightning and chaining it to a machine.

The keeper of this coil is a guy named Jaime. He is small, somewhat like a hobbit, and exactly the kind of tinkerer/geek who builds enormous Tesla coils in his free time. (Alex found him through an Internet chat board.) Jaime scurries around, checking connections and muttering. It seems as though exposure to high-intensity electrical fields has had a deleterious effect on his sociability. Ask him a question and he moves his lips silently for a second or two before and after answering, as though he is muttering dark qualifiers.

"Yes, the field is safe," he says. And then his lips keep moving, as if to say, Safe for me. But it might melt your face.

I'm worried because I've heard disquieting stories about the unpredictability of coils. Alex told me that while Googling he'd found an account of a recent accident with this very apparatus. The scene: a Burning Man–esque piece in which the coil fired lightning while a woman danced nearby holding a fluorescent tube (a live Tesla coil pumps so much electricity into the air that fluorescent bulbs spontaneously ignite, which looks extremely cool). An errant charge leapt from the coil to the fluorescent tube and then grounded itself on her nipple ring.

"Mike, do you have any metal in your body?" Alex asks.

"I don't think so," I say.

"Mike," he says, very deliberately. "You'd better know whether you have any metal in your body."

By now I'm sure that my body is metal free, but I'm still nervous as Jaime flips two switches and then turns an enormous knob. A high-pitched schreeak fills the room, like sheet metal getting torn. Lightning pours out of the toroid and writhes around like a cat on a short leash. Arcs leap out and dance across chairs, the ceiling, and the floor.

Jaime gestures for me to get closer to the coil. As I do so nervously, he crouches lower behind the control box, continuously adjusting that one knob with the intensity of a Trappist monk. He is very proud of the coil and touches it constantly as he tells us he will need a great deal of money to bring it to performances. When we regretfully decline, he abruptly reverses and says he can do it for free, so long as we provide him with dinner. This sounds too good to be true, and, in fact, it is. A few days later, he changes his mind again and calls demanding compensation.

It's a lot of money, but I've decided to fight for the coil. I've become enchanted by it. I want to be the mad scientist, bringing a terrifying and glorious new machine out in front of the masses. Just once, I want to shout, "Bwahahhaha-haha!" and really mean it.

Just a day before the opening, Alex the artistic director tells me that the designers and technicians have demanded a meeting to discuss the coil. He explains that the designers and technicians are totally freaked out—they fear it will blow out all of their electronics and who knows what else. Tech has already been going poorly—the live rabbits are dying of an unknown ailment, and the multimedia is on the fritz. The last thing anyone wants is a lightning-throwing death machine.

We all meet in the theater: designers, technicians, me, and Jaime, the Keeper of the Coil. Alex tosses out what should be a softball question: "Jaime, what's the worst thing that could happen with the coil?" I turn expectantly, eager to hear his impassioned defense. He thinks for a moment, his lips moving, and then says, slowly, "Well . . . it could kill someone."

The designers are horrified—they were worried about the multimedia getting ruined, and now we're talking about murdering audience members. "Jaime," I say, hurriedly spinning his answer, "what you mean by that is that the coil could kill someone if they doused themselves in water and jumped on top of it, right, which is why we have a safety zone, right?"

"Yes, that's true," he responds. "And it's also true that sometimes electricity does whatever it wants. It's hard to predict."

The theater management hears about our meeting and wants us to get supplemental insurance that covers lightning-throwing death machines. The Tesla coil is axed. It sits offstage during the performances, disassembled and inert in a paint closet. It's the loveliest, most frightening, and saddest machine I have ever seen.

Daniel H. Pink

The Book Stops Here

Jimmy Wales wanted to build a free encyclopedia on the Internet. So he raised an army of amateurs and created the self-organizing, self-repairing, hyperaddictive library of the future called Wikipedia.

Dixon, New Mexico, is a rural town with a few hundred residents and no traffic lights. At the end of a dirt road, in the shadow of a small mountain, sits a gray trailer. It is the home of Einar Kvaran. To understand the most audacious experiment of the postboom Internet, this is a good place to begin.

Kvaran is a tall and hale 56-year-old with a ruddy face, blue eyes, and blond hair that's turning white. He calls himself an "art historian without portfolio" but has no formal credentials in his area of proclaimed expertise. He's never published a scholarly article or taught a college course. Over three decades, he's been a Peace Corps volunteer, an autoworker, a union steward, a homeschooling mentor, and the drummer in a Michigan band called Kodai Road. Right now, he's unemployed. Which isn't to say he doesn't work. For about six hours each day, Kvaran reads and writes about

American sculpture and public art and publishes his articles for an audience of millions around the world.

Hundreds of books on sculptors, regional architecture, and art history are stacked floor to ceiling inside his trailer—along with 68 thick albums containing 20 years of photos he's taken on the American road. The outlet for his knowledge is at the other end of his dial-up Internet connection: the daring but controversial Web site known as Wikipedia.

Four years ago, a wealthy options trader named Jimmy Wales set out to build a massive online encyclopedia ambitious in purpose and unique in design. This encyclopedia would be freely available to anyone. And it would be created not by paid experts and editors but by whoever wanted to contribute. With software called Wiki—which allows anybody with Web access to go to a site and edit, delete, or add to what's there—Wales and his volunteer crew would construct a repository of knowledge to rival the ancient library of Alexandria.

In 2001, the idea seemed preposterous. In 2005, the nonprofit venture is the largest encyclopedia on the planet. Wikipedia offers 500,000 articles in English—compared with *Britannica*'s 80,000 and *Encarta*'s 4,500—fashioned by more than 16,000 contributors. Tack on the editions in 75 other languages, including Esperanto and Kurdish, and the total Wikipedia article count tops 1.3 million.

Wikipedia's explosive growth is due to the contributions of Kvaran and others like him. Self-taught and self-motivated, Kvaran wrote his first article last summer—a short piece on American sculptor Corrado Parducci. Since then, Kvaran has written or contributed to two dozen other entries on American art, using his library and photographs as sources. He's added words and images to 30 other topics, too—the Lincoln Memorial; baseball player Carl Yastrzemski; photographer Tina Modotti; and Iceland's first prime minister, Hannes Hafstein, who happens to be Kvaran's great-grandfather. "I think of myself as a teacher," Kvaran says over tea at his kitchen table.

To many guardians of the knowledge cathedral—librarians, lexicographers, academics—that's precisely the problem. Who died and made this guy professor? No pedigreed scholars scrutinize his work. No research assistants check his facts. Should we trust an encyclopedia that allows anyone with a pulse and a mouse pad to opine about Jackson Pollock's place in postmodernism? What's more, the software that made Wikipedia so easy to build also makes it easy to manipulate and deface. A former editor at the venerable *Encyclopædia Britannica* recently likened the site to a public restroom: You never know who used it last.

So the modest trailer at the end of a dirt road in this pinprick of a town holds some cosmic questions. Is Wikipedia a heartening effort in digital humanitarianism—or a not-so-smart mob unleashing misinformation on the masses? Are well-intentioned amateurs any replacement for professionals? And is charging nothing for knowledge too high a price?

Recovery may take 12 steps, but becoming a junkie requires only 4. First comes chance—an unexpected encounter. Chance stirs curiosity. Curiosity leads to experimentation. And experimentation cascades into addiction.

For Danny Wool, chance arrived on a winter afternoon in 2002, after an argument about—of all things—Kryptonite. Googling the term from his Brooklyn home to settle the debate, he came upon the Wikipedia entry. He looked up a few more subjects and noticed that each one contained a mysterious hyperlink that said Edit. Curious but too nervous to do anything, he returned to Wikipedia a few more times. Then one night he corrected an error in an article about Jewish holidays. *You can do that?!* It was his first inhalation of Wiki crack. He became one of Wikipedia's earliest registered users and wrote his first article—on Muckleshoot, a Washington State Indian tribe. Since then, he has made more than 16,000 contributions.

Bryan Derksen wrote the original Kryptonite article that Wool discovered. While surfing from his home in Edmonton, Derksen also stumbled upon Wikipedia and quickly traveled the path to addiction. He read a few entries on Greek mythology and found them inadequate. The Edit link beckoned him like a street pusher. He clicked it and typed in a few changes. *You can do that?!* "I just got hooked," he tells me. He's now made more edits than all but three Wikipedians—some 40,000 additions and revisions.

Number one on the list of contributors is Derek Ramsey, who has automated his addiction. A software engineer in Pennsylvania, Ramsey wrote a Java program called rambot that automatically updates Wikipedia articles on cities and counties. So far, the man and machine combination has contributed more than 100,000 edits.

String enough of these addicts together, add a few thousand casual users, and pretty soon you have a new way to do an old thing. Humankind has long sought to tame the jungle of knowledge and display it in a zoo of friendly facts. But while the urge to create encyclopedias has endured, the production model has evolved. Wikipedia is the latest stage.

In the beginning, encyclopedias relied on the One Smart Guy model. In ancient Greece, Aristotle put pen to papyrus and single-handedly tried to record all the knowledge of his time. Four hundred years later, the Roman nobleman Pliny the Elder cranked out a 37-volume set of the day's knowledge. The Chinese scholar Tu Yu wrote an encyclopedia in the ninth century. And in the 1700s, Diderot and a few pals (including Voltaire and Rousseau) took 29 years to create the *Encyclopædie, ou Dictionnaire Raisonné des Sciences, des Arts et des Métiers.*

With the Industrial Revolution, the One Smart Guy approach gradually gave way to the One Best Way model, which borrowed the principles of scientific management and the lessons of assembly lines. *Encyclopædia Britannica* pioneered this approach in Scotland and honed it to perfection. Large groups of experts, each performing a task on a detailed work chart under the direction of a manager, produced encyclopedias of enormous breadth. Late in the 20th century, computers changed encyclopedias—and the Internet changed them more. Today, *Britannica* and *World Book* still sell some 130-pound, $1,100, multivolume sets, but they earn most of their money from Internet subscriptions. Yet while the medium has shifted from atoms to bits, the production model—and therefore the product itself—has remained the same.

Now Wales has brought forth a third model—call it One for All. Instead of one really smart guy, Wikipedia draws on thousands of fairly smart guys and gals—because in the metamathematics of encyclopedias, 500 Kvarans equals one Pliny the Elder. Instead of clearly delineated lines of authority, Wikipedia depends on radical decentralization and self-organization—open source in its purest form. Most encyclopedias start to fossilize the moment they're printed on a page. But add Wiki software and some helping hands and you get something self-repairing and almost alive. A different production model creates a product that's fluid, fast, fixable, and free.

The One for All model has delivered solid results in a remarkably short time. Look up any topic you know something about—from the Battle of Fredericksburg to *Madame Bovary* to Planck's law of blackbody radiation—and you'll probably find that the Wikipedia entry is, if not perfect, not bad. Sure, the Leonard Nimoy entry is longer than the one on Toni Morrison. But the Morrison article covers the basics of her life and literary works about as well as the *World Book* entry. And among the nearly half million articles are tens of thousands whose quality easily rivals that of *Britannica* or *Encarta.*

What makes the model work is not only the collective knowledge and effort of a far-flung labor force but also the willingness to abide by two core principles. The first: neutrality. All articles should be written without bias. Wikipedians are directed not to take a stand on controversial subjects like abortion or global warming but to fairly represent all sides. The second principle is good faith. All work should be approached with the assumption that the author is trying to help the project, not harm it.

Wikipedia represents a belief in the supremacy of reason and the goodness of others. In the Wikipedia ideal, people of goodwill sometimes disagree. But from the respectful clash of opposing viewpoints and the combined wisdom of the many, something resembling the truth will emerge. Most of the time.

If you looked up Jimmy Carter on Wikipedia one morning this winter, you would have discovered something you couldn't learn from *Britannica.* According to the photo that accompanied Carter's entry, America's 39th president was a scruffy, random, unshaven man with his left index finger shoved firmly up his nose.

Lurking in the underbrush of Wikipedia's idyllic forest of reason and good intentions are contributors less noble in purpose, whose numbers are multiplying. Wiki devotees have names for many of them. First, there are the trolls, minor troublemakers who breach the principle of good faith with inane edits designed to rile serious users. More insidious are vandals, who try to wreck the site—inserting profanity and ethnic slurs, unleashing bots that put ads into entries, and pasting pictures of penises and other junior-high laugh getters. Considering how easy it is to make changes on Wikipedia, you'd imagine these ne'er-do-wells could potentially overwhelm the site. But they haven't—at least not yet—because defenses against them are built into the structure.

Anybody who is logged in can place an article on a "watch list." Whenever somebody amends the entry, the watch list records the change. So when that anonymous vandal replaced a Jimmy Carter photo with a nose picker, all the Wikipedians with Jimmy Carter on their watch list knew about it. One of them merely reverted to the original portrait. At the same time, the user who rescued the former president from Boogerville noticed that the vandal had also posted the nose-pick photo on the "Rapping" entry—and he got rid of that image just four minutes after the photo appeared.

On controversial topics, the response can be especially swift. Wikipedia's article on Islam has been a persistent target of vandalism, but Wikipedia's defenders of Islam have always proved nimbler than the vandals. Take one fairly typical instance. At 11:20 one morning not too long ago, an anonymous user replaced the entire Islam entry with a single scatological word. At 11:22, a user named Solitude reverted the entry. At 11:25, the anonymous user struck again, this time replacing the article with the phrase "u stink!" By 11:26, another user, Ahoerstemeir, reverted that change—and the vandal disappeared. When MIT's Fernanda Viégas and IBM's Martin Wattenberg and Kushal Dave studied Wikipedia, they found that cases of mass deletions, a common form of vandalism, were corrected in a median time of 2.8 minutes. When an obscenity accompanied the mass deletion, the median time dropped to 1.7 minutes.

It turns out that Wikipedia has an innate capacity to heal itself. As a result, woefully outnumbered vandals often give up and leave. (To paraphrase Linus Torvalds, given enough eyeballs, all thugs are callow.) What's more, making changes is so simple that who prevails often comes down to who cares more. And hardcore Wikipedians care. A lot.

Wool logs on to Wikipedia at 6 each morning and works two hours before leaving for his day job developing education programs for a museum. When he gets back home around 6:30 p.m., he hops back on Wikipedia for a few more hours. Derksen checks his watch list each morning before leaving for work at a small company that sells medical equipment on eBay. When he returns home, he'll spend a few hours just clicking on the Random Page link to see what needs to get done. It's tempting to urge people like Wool and Derksen to get a life. But imagine if they instead spent their free time walking through public parks, picking up garbage. We'd call them good citizens.

Still, even committed citizens sometimes aren't muscular enough to fend off determined bad guys. As Wikipedia has grown, Wales has been forced to impose some more centralized, policelike measures—to guard against "edit warriors," "point-of-view warriors," "revert warriors," and all those who have difficulty playing well with others. "We try to be as open as we can," Wales says, "but some of these people are just impossible." During the last presidential election, Wikipedia had to lock both the George W. Bush and the John Kerry pages because of incessant vandalism and bickering. The Wikipedia front page, another target of attacks, is also protected.

If that suggests an emerging hierarchy in this bastion of

egalitarian knowledge gathering, so be it. The Wikipedia power pyramid looks like this: At the bottom are anonymous contributors, people who make a few edits and are identified only by their IP addresses. On the next level stand Wikipedia's myriad registered users around the globe, people such as Kvaran in New Mexico, who have chosen a screen name (he's Carptrash) and make edits under that byline. Some of the most dedicated users try to reach the next level—administrator. Wikipedia's 400 administrators, Derksen and Wool among them, can delete articles, protect pages, and block IP addresses. Above this group are bureaucrats, who can crown administrators. The most privileged bureaucrats are stewards. And above stewards are developers, 57 superelites who can make direct changes to the Wikipedia software and database. There's also an arbitration committee that hears disputes and can ban bad users.

At the very top, with powers that range far beyond those of any mere Wikipedian mortal, is Wales, known to everyone in Wiki world as Jimbo. He can do pretty much anything he wants—from locking pages to banning people to getting rid of developers. So vast are his powers that some began calling him "the benevolent dictator." But Wales bristled at that tag. So his minions assigned him a different, though no less imposing, label. "Jimbo," says Wikipedia administrator Mark Pellegrini, "is the God-King."

The God-King drives a Hyundai. On a sunny Florida Monday, Wales is piloting his red Accent from his St. Petersburg home across the bay to downtown Tampa, where on the 11th floor of a shabby office building a company called Neutelligent manages a vast server farm. In one of the back rows, stacked on two racks, are the guts of Wikipedia—42 servers connected by a hairball of orange and blue cables. For the next two hours, Wales scoots to and fro, plugging and unplugging cables while trading messages with a Wikipedia developer on Internet Relay Chat via a nearby keyboard.

Back in St. Pete, Wales oversees his empire from a pair of monitors in Wikipedia's headquarters—two cramped, windowless rooms that look like the offices of a failed tech startup. Computer equipment is strewn everywhere. An open copy of *Teach Yourself PHP, MySQL, and Apache* is splayed on the floor. It may be good to be God-King, but it's not glamorous.

Wales began his journey in Huntsville, Alabama. His father worked in a grocery store. His mother and grandmother operated a tiny private school called the House of Learning, which Wales and his three siblings attended. He graduated from Auburn University in 1989 with a degree in finance and ended up studying options pricing in an economics Ph.D. program at Indiana University. Bored with academic life, he left school in 1994 and went to Chicago, where he took to betting on interest rate and foreign currency fluctuations. In six years, he earned enough to support himself and his wife for the rest of their lives.

They moved to San Diego in 1998. The times being what they were, Wales started an Internet company called Bomis, a search engine and Web directory. He began hearing about the fledgling open source movement and wondered whether volunteers could create something besides software. So he recruited Larry Sanger, then an Ohio State University doctoral student in philosophy, whom he'd encountered on some listservs. He put Sanger on the Bomis payroll, and together they launched a free online encyclopedia called Nupedia. Why an encyclopedia? Wales says he simply wanted to see if it could be done.

With Sanger as editor in chief, Nupedia essentially replicated the One Best Way model. He assembled a roster of academics to write articles. (Participants even had to fax in their degrees as proof of their expertise.) And he established a seven-stage process of editing, fact-checking, and peer review. "After 18 months and $250,000," Wales says, "we had 12 articles."

Then an employee told Wales about Wiki software. On January 15, 2001, they launched a Wiki-fied version and within a month, they had 200 articles. In a year, they had 18,000. And on September 20, 2004, when the Hebrew edition added an article on Kazakhstan's flag, Wikipedia had its one millionth article. Total investment: about $500,000, most of it from Wales himself.

Sanger left the project in 2002. "In the Nupedia model, there was room for an editor in chief," Wales says. "The Wiki model is too distributed for that." Sanger, a scholar at heart, returned to academic life. His cofounder, meanwhile, became a minor geek rock star. Wales has been asked to advise the BBC, Nokia, and other large enterprises curious about Wikis. Technology conferences in the United States and Europe clamor for him. And while he's committed to keeping his creation a "charitable project," as he constantly calls it (wikipedia.com became wikipedia.org almost three years ago), the temptations are mounting.

Late last year, Wales and Angela Beesley, an astonishingly dedicated Wikipedian, launched a for-profit venture called WikiCities. The company will provide free hosting for "community-based" sites—RVers, poodle owners, genealogy buffs, and so on. The sites will operate on the same software that powers Wikipedia, and the content will be available under a free license. But WikiCities intends to make money by selling advertising. After all, if several thousand people can create an encyclopedia, a few hundred Usher devotees should be able to put together the ultimate fan site. And if legions of Usher fans are hanging out in one place, some advertiser will pay to try to sell them concert tickets or music downloads.

It may feel like we've been down this road before—remember GeoCities and theglobe.com? But Wales says this is different because those earlier sites lacked any mechanism for true community. "It was just free home pages," he says. WikiCities, he believes, will let people who share a passion also share a project. They'll be able to design and build projects together. So the founder of the Web's grand experiment in the democratic dissemination of information is also trying to resurrect GeoCities. While some may find the notion silly, many others just want a piece of Jimbo magic.

During our conversation over lunch, Wales's cell phone rings. It's a partner at Accel, the venture capital firm, calling to talk about WikiCities and any other Wiki-related investment ideas Wales might have. Wales says he's busy and asks the caller to phone back later. Then he smiles at me. "I'll let him cool his heels awhile."

Wikipedia's articles on the British peerage system—clear-headed explanations of dukes, viscounts, and other titles of nobility—are largely the work of a user known as Lord Emsworth. A few of Emsworth's pieces on kings and queens of England have been honored as Wikipedia's Featured Article of the Day. It turns out that Lord Emsworth claims to be a 16-year-old living in South Brunswick, New Jersey. On Wikipedia, nobody has to know you're a sophomore.

And that has some distressed. Larry Sanger gave voice to these criticisms in a recent essay posted on kuro5hin.org titled "Why Wikipedia Must Jettison Its Anti-Elitism." Although he acknowledges that "Wikipedia is very cool," he argues that the site's production model suffers from two big problems.

The first is that "regardless of whether Wikipedia actually is more or less reliable than the average encyclopedia," librarians, teachers, and academics don't perceive it as credible because it has no formal review process. The second problem, according to Sanger, is that the site in general and Wales in particular are too "anti-elitist." Established scholars might be willing to contribute to Wikipedia—but not if they have to deal with trolls and especially not if they're considered no different from any schmo with an iMac.

Speaking from his home in Columbus, Ohio, where he teaches at Ohio State, Sanger stresses that Wikipedia is a fine and worthy endeavor. But he says that academics don't take it seriously. "A lot of the articles look like they're written by undergraduates." He believes that "people who make knowing things their life's work should be accorded a special place in the project." But since Wikipedia's resolute antielitism makes that unlikely, Sanger argues, something else will happen: Wikipedia will fork—that is, a group of academics will take Wikipedia's content, which is available under a free license, and produce their own peer-reviewed reference work. "I wanted to send a wake-up call to the Wikipedia community to tell them that this fork is probably going to happen."

Wales' response essentially boils down to this: Fork you. "You want to organize that?" he sniffs. "Here are the servers." Yet Wales acknowledges that in the next year, partly in response to these concerns, Wikipedia will likely offer a stable—that is, unchangeable—version alongside its One for All edition.

But both Sanger's critique and Wales' reaction miss a larger point: You can't evaluate Wikipedia by traditional encyclopedia standards. A forked Wikipedia run by academics would be Nupedia 2.0. It would use the One Best Way production model, which inevitably would produce a

One Best Way product. That's not a better or worse Wikipedia any more than Instapundit.com is a better or worse *Washington Post.* They are different animals.

Encyclopedias aspire to be infallible. But Wikipedia requires that the perfect never be the enemy of the good. Citizen-editors don't need to make an entry flawless. They just need to make it better. As a result, even many Wikipedians believe the site is not as good as traditional encyclopedias. "On a scale of 1 to 10, I'd give Wikipedia a 7.8 in reliability," Kvaran told me in New Mexico. "I'd give *Britannica* an 8.8." But how much does that matter? *Britannica* has been around since before the American Revolution; Wikipedia just celebrated its fifth birthday. More important, *Britannica* costs $70 a year; Wikipedia is free. The better criterion on which to measure Wikipedia is whether this very young, pretty good, ever improving, totally free site serves a need—just as the way to measure *Britannica* is whether the additional surety that comes from its production model is worth the cost.

There's another equally important difference between the two offerings. The One Best Way approach creates something finished. The One for All model creates something alive. When the Indian Ocean tsunami erupted late last year, Wikipedians produced several entries on the topic within hours. By contrast, World Book, whose CD-ROM allows owners to download regular updates, hadn't updated its tsunami or Indian Ocean entries a full month after the devastation occurred. That's the likely fate of Wikipedia's proposed stable, or snapshot, version. Fixing its contents in a book or on a CD or DVD is tantamount to embalming a living thing. The body may look great, but it's no longer breathing.

"You can create life in there," says Wikipedian Oliver Brown, a high school teacher in Aptos, California. "If you don't know about something, you can start an article, and other people can come and feed it, nurture it." For example, two years ago, Danny Wool was curious about the American architectural sculptor Lee Lawrie, whose statue of Atlas sits nearby Rockefeller Center. Wool posted a stub—a few sentences on a topic—in the hopes that someone would add to it. That someone turned out to be Kvaran, who owned several books on Lawrie and who'd photographed his work not only at Rockefeller Center but also at the Capitol Building in Lincoln, Nebraska. Today, the Lawrie entry has grown from two sentences to several thorough paragraphs, a dozen photos, and a list of references. Brown himself posted a stub when he was wondering how many people were considered the father or mother of something. Today Wikipedia lists more than 230 people known as the father or mother of an idea, a movement, or an invention. And that number will likely be higher tomorrow. As the father of this new kind of encyclopedia puts it, "Wikipedia will never be finished."

In 1962, Charles Van Doren—who would go on to become a senior editor of *Britannica* but is more famous for his role in the 1950s quiz show scandal—wrote a think piece for the journal the *American Behavioral Scientist.* His essay, "The Idea of an Encyclopedia," is similar in spirit to the one Sanger wrote late last year: a warning to his community.

Van Doren warned not that encyclopedias of his day lacked credibility but that they lacked vitality. "The tone of American encyclopedias is often fiercely inhuman," he wrote. "It appears to be the wish of some contributors to write about living institutions as if they were pickled frogs, outstretched upon a dissecting board." An encyclopedia ought to be a "revolutionary document," he argued. And while Van Doren didn't call for a new production model, he did say that "the ideal encyclopedia should be radical. It should stop being safe."

What stood in the way of this new approach was precisely what encyclopedias prided themselves on. "Respectability seems safe," he wrote. "But what will be respectable in 30 years seems avant-garde now. If an encyclopedia hopes to be respectable in 2000, it must appear daring in the year 1963."

Jimbo and his minions—from Einar Kvaran in his New Mexico trailer to Lord Emsworth in his New Jersey bedroom—may seem daring today. But they're about to become respectable.

Joseph Turow

Have They Got a Deal for You

It's suspiciously cozy in the cybermarket.

A couple of years ago, in an undergraduate seminar I taught called "Spam and Society," discussion veered a bit off topic. One of the students asserted confidently that airline Web sites give first-time users lower prices than returning customers. Most of the others immediately agreed. They said the motive was to suck in potential buyers; then, when they returned, the airline could quietly raise prices.

I hear this kind of claim fairly often among heavy computer users. It seems to have become an article of faith that the unseen moguls behind all sorts of Web sites are cherry-picking consumers, customizing ads, manipulating prices, and changing product offers based on what they've learned about individual users without the users' knowledge.

In my research on Internet marketing, I talk to lots of Web workers and consultants and read the trade press, and it's pretty clear that this is going on. But it's extraordinarily hard to verify when it occurs, why any particular offer is made, or how a vendor is evaluating any given customer. Online merchants don't have to tell anyone how they operate, so generally they don't. University of Utah professor Rob Mayer, an adviser to Consumers Union's WebWatch

/*Washington Post*/

project, told me he once asked an online travel industry executive whether travel sites offer certain customers different prices depending on their online history. The reply was cryptic, yet telling: "I won't say it doesn't happen."

Lighten up, you might say. Nobody's forcing anyone to buy airline tickets or anything else. But I'm disturbed by what this reflects about our general retail environment—the evolution of what I would call a culture of suspicion. From airlines to supermarkets, from banks to Web sites, American consumers increasingly believe they are being spied on and manipulated. But they continue to trade in the marketplace because they feel powerless to do anything about it.

This is a profound change. Broadly speaking, the past 150 years saw what you might call the democratization of shopping in the United States. Beginning around the mid-1800s, department stores such as Stewart's in New York City and Wanamaker's in Philadelphia moved away from the haggling mode of selling and began to display goods and uniform prices for all to see. Part of the motive was self-interest: Given the wide variety of merchandise and the large number of employees, the store owners didn't trust their clerks to bargain well with customers. But the result was a more or less egalitarian, transparent marketplace—one that Americans have come to take for granted. When they have to negotiate—most notably in car dealerships—they see it as an unusual, nerve-racking experience.

This reliance on evenhanded, fair dealing lies at the heart of American capitalism or at least the way we'd like to think of it. It's not always practiced, by any means, as antitrust suits and many consumer complaints attest. But it's a worthy goal, and such institutions as the Securities and Exchange Commission and the Federal Trade Commission were established, in part, to aim for it.

The scaffolding of this system is shaken if a retailer

changes its offerings to individual consumers based on information about the consumers that the consumers don't know or that they suspect but can't verify. Take airline tickets. The Consumer Union's WebWatch investigators visited airline sites many hundreds of times, finding a bewildering array of prices that seemed weirdly random and virtually unpredictable (some prices changed between log-on and checkout). The airlines say it's the result of a necessarily complex pricing structure, but how can we tell that's all that's going on?

In September 2000, Amazon.com got headlines when customers found that the same DVDs were being offered to different buyers at discounts of 30, 35, or 40 percent. Amazon insisted the discounts were part of a random "price test," but critics suggested they were based on customer profiling. After weeks of bad press, the firm offered to refund the difference to buyers who had paid the higher prices. The company vowed it wouldn't happen again.

Frequent computer users I've talked to—like my fairly hip students—don't really believe such assurances. Frankly, it's hard for any dispassionate observer to believe there's no "price customization" when associates from the influential McKinsey consulting firm write in a 2004 *Harvard Business Review* article that online companies are missing out on a "big opportunity" if they are not tracking customers and adjusting prices accordingly—either to attract new buyers or get more of their money.

Meanwhile, this sort of thing goes on quite openly in brick-and-mortar retail stores. In a hypercompetitive environment, where trying to beat Wal-Mart and Costco on price is all but impossible, department stores and supermarkets compete with them by trying to hook the right customers. Operating on the financial industry's premise that about 20 percent of the customers bring in 80 percent of profits, they try to identify who belongs in that 20 percent—who will spend money and come back to spend more. And they're happy to get rid of those who hold out for bargains or return too many purchases.

Note that this is subtly different from the 20th-century model, in which storekeepers stocked certain brands of soap or pasta or disposable diapers because they sold well, thus pleasing customers and making money at the same time. While that certainly continues, the new goal is to make money by identifying individuals who fit "best customer" profiles and then reinforce their purchases for reasons and in ways that are hidden from them.

It happens everywhere. Many banks give customers scores based on their deposits and financial backgrounds; their phone representatives use friendlier scripts for high scorers. Supermarket cash registers spit out customized coupons at checkouts, tailored to the previous purchases recorded on "preferred customer" cards. If you buy Coke, you might get a coupon for Pepsi—or perhaps one for Coke, to make sure you return; if you buy coffee, they'll offer you a discount latte. Some grocery chains are already testing "shopping buddies," small computers that you actually carry around the store, getting personally tailored recommendations and discounts from the moment you enter.

The idea used to be that you, the consumer, could shop around, compare goods and prices, and make a smart choice. But now the reverse is also true: The vendor looks at its consumer base, gathers information, and decides whether you are worth pleasing or whether it can profit from your loyalty and habits. You may try to jump from site to site to hunt for the best buy, but that's time consuming. And there are comparative shopping sites such as Bizrate or Nextag, but these can be tough to navigate and companies are learning quickly how to game the system.

This all might make sense for retailers. But for the rest of us, it can feel like our simple corner store is turning into a Marrakech bazaar—except that the merchant has been reading our diary while we're negotiating blindfolded, behind a curtain, through a translator.

Considering the manic nature of retail and media competition and technological developments, this kind of marketing is likely to grow and become more sophisticated. I've spoken to telemarketing executives who claim they can target TV commercials to specific households. It is even possible, they say, to digitally place different products or dialogue into the scene of a show. Sure, there are financial and technical roadblocks to implementing these customized commercials—not to mention some charmingly old-fashioned concerns about privacy—but I'll bet they're coming.

Certainly, being targeted by marketers has its benefits. I like getting coupons for my favorite breakfast cereal. I like Amazon.com suggesting films I might enjoy. And I like being treated well on the phone by a company that thinks I'm valuable. But I also like feeling that I'm in control of buying the product, not that the seller is choosing me.

In the early 20th century, "keeping up with the Joneses" was a real ideal, and it spurred consumption. But the mysteries surrounding database marketing will increasingly make us not so much competitive as wary: Are our neighbors getting a better deal not because they shopped harder or bargained smarter but because of some database demographic we don't know about and can't fight?

Lack of knowledge breeds suspicion. A survey I directed this year for the Annenberg Public Policy Center found a startling degree of high-tech ignorance among Americans who use the Internet. Eighty percent of those interviewed knew that companies can track their activities across the Web. Yet a substantial majority believe, too, that it is illegal for merchants and charities to sell information about them, even though it's legal and goes on all the time. Sixty-four percent believe incorrectly that "a site such as Expedia or Orbitz that compares prices on different airlines must include the lowest airline prices." And only 25 percent knew that the following statement was false: "When a website has a privacy policy, it means the site will not share my information with other websites and companies." Our report calls for changing the Orwellian label "Privacy policy" to the more honest "Using your information."

We also call for merchants to be more open about their database activities. A friend of mine phoned Citibank about his credit card and got a representative from India who wasn't able to understand his problem. My friend had never before gotten an "outsourced" operator and, knowing something about databases, concluded that Citibank had pushed him down a notch in status, reserving the U.S.-based service people for better customers. The reality is, he knows nothing about how Citibank operates and has no idea what it thinks about his status. My point is, precisely because he's Internet savvy, he's automatically suspicious that information may be used against him without his knowing it.

We used to say, "My money's as good as anyone else's." But in the 21st century, that may no longer be true. My students can hardly remember a time before Internet cookies and frequent flying and preferred shopping. They and their kids will try to beat the system to get the best deals, all the while assuming that they don't know all the rules of the marketplace. They'll be automatically mistrustful. It's a new world out there.

Steven Levy

The Trend Spotter

Tim O'Reilly built an empire on computer manuals and conferences that make sense of new technologies. Along the way he became the guru of the participation age.

Tim O'Reilly likes to walk in the dark. Sometimes after dinner he'll head down the long dirt pathway outside his rambling farmhouse in Sebastopol, California, a posthippie enclave between wine country and the Pacific Ocean. No flashlight. And on this particular overcast night, with rain dropping from a mossy sky, it's tough to see a thing. But O'Reilly's pace is brisk and optimistic, his feet crunching the dirt as the wiry 51-year-old hurtles himself forward.

As always, he's relying on his radar for safe passage. O'Reilly's radar is legendary. It works on country roads and on the information sea. It told him there was a market for consumer-friendly computer manuals and that he could build a great business publishing them. It helped him understand the significance of the World Wide Web before there were browsers to surf it. And it led him to identify and proselytize technologies like peer-to-peer, syndication, and WiFi before most people had even heard of them at all. As a result, "Tim O'Reilly's radar" is kind of a catchphrase in the industry.

Yet O'Reilly himself has operated for years *under* the radar. Most nontechies, if they know him at all, know him by the eponymous name of his publishing company. It has a 15 percent share of the $400 million computer-book market but casts a much bigger shadow. O'Reilly books tend to colonize entire sections at Borders and Barnes & Noble, their distinctive cover design as recognizable as the Tide circle on a box of detergent or the Apple logo on the lid of a PowerBook. In serif type over a glossy white background, there is the title, often naming a computer language or protocol familiar to codeheads and gibberish to everyone else (*JavaServer Faces; Essential CVS; Using Samba, 2nd Edition*). The illustrations are realistically rendered pen-and-ink drawings of animals.

Now the man behind the menagerie is emerging as a figure in his own right. The conferences he hosts generate a lot of buzz; they also establish a sort of geek's conventional wisdom and build a community around O'Reilly's own hand-rolled view of what's hot or going to be hot. His conclaves come in varying shapes and sizes, from large gatherings like Emerging Technology (ETech)—for which programmers, entrepreneurs, and venture capitalists pack hotel conference rooms every spring—to the intimate Foo (Friends of O'Reilly) camps, which are like pajama parties for propeller heads. No matter the event, the highlight is usually O'Reilly's opening talk. Hair a bit mussed, eyes gleaming behind wire-frame spectacles, dressed in studiously informal garb—khakis topped with T-shirt or fleece—O'Reilly will insist to his audience that "you are our radar; we're just picking it up."

Most recently, O'Reilly's dead-on inner compass led

him to anticipate the current stage of the Internet. Powered by the bottom-up nature of sharing and collective action, it's exemplified by such developments as the barn-raising methodology of Wikipedia; group efforts like tagging; open source systems; Wi-Fi; open application programming interfaces (APIs) in e-commerce sites like Amazon, eBay, and Google; really simple syndication (RSS); the spontaneous connectivity of Apple's Rendezvous; and dozens of other dots that are being connected to fulfill the original promise of the Net. He calls it the architecture of participation. In O'Reilly's world, sharing increases value—so much so that it becomes unthinkable to close off information or to adopt nonstandard proprietary systems. The result is a virtuous cycle where openness becomes the norm, encouraging even more participation.

I got my first O'Reilly briefing on the paradigm of participation last year, at one of his ETech conferences. "We're really at the stage where the Net is the center of everything," he said. "The rules are different. We have a new shift in power." I pointed out that previous ecstatic views of the Internet had been buried under waves of spam, fraud, regulation, and the dot-com disaster. But O'Reilly remains optimistic. "I like to think that even if we make some really bad choices and go down some bad paths, we'll eventually emerge from it."

Then he took a conversational turn that you seldom hear from your garden-variety tech analyst. "There's a wonderful phrase in a different context from Wallace Stevens, talking about a metaphysical view of reality versus the information." From memory, he translated a passage from Stevens's "Esthétique du Mal." The key phrase: "Beneath every no / Lay a passion for yes that had never been broken."

"I just love that," O'Reilly said, his ever-present sheepish grin taking on a late-afternoon wistfulness. "'Beneath every no lay a yes that had never been broken.' To me, there's this wonderful *spirit*. . . . You know, I believe people are fundamentally good and want to find things that make life better for themselves. There are social dynamics for people that work, and there are ones that are pathological. But beneath every no lay a yes that had never been broken. I put my life-faith in that."

And to think we were just talking about Internet protocols. O'Reilly's theories about the next Internet seem on the mark. But the impromptu poetry recital diverted my attention to O'Reilly himself. As it turns out, the levers and pulleys of this new Net neatly reflect the operating principles of the man who helped define it: a philosophy of participation and sharing and a sense that collective action will inevitably accrue to the greater good.

The crucial technologies that make this happen—the digital infrastructure that makes the online world a perpetual swap meet of goods and ideas—are the culmination of all the stuff he's been tracking, supporting, and popularizing for the past 20 years. After O'Reilly hosted his Web 2.0 Conference last autumn (an even bigger confab is planned for this October), digi-pundits began using the term Web 2.0 to describe the growing swarm of applications and Web sites that exploit collective intelligence and participation. But they may as well call it Tim O'Reilly's Internet.

O'Reilly's ability to describe trends in terms of human comity springs from an even deeper source. To find it, I take the road to laid-back Sebastopol, 50 miles north of San Francisco and 35 years back into the cultural past. (Signs at the town line proudly proclaim Sebastopol a nuclear-free zone.) This is the end point of a journey that began long ago for O'Reilly: an epic walk in the dark, if you will, that led to a clearing along the road.

I tour two of the three chalet-style buildings that house

his company and have dinner with O'Reilly and his wife, Christina, a vivacious theater person (she is starting her own company, called Light Touch Theater). Then I follow him down some rural roads to his home, set on six acres of farmland. Before I retire to the guest quarters, on the second floor of a garagelike building across a dirt path, he shows me his vast book collection, which stretches across several rooms of his rambling house. I pick up one of the strangest volumes I've ever seen: a dense edition of the journals of George Simon, a little-known teacher of consciousness studies. I borrow the book because it has been painstakingly compiled by Simon's most faithful acolyte—Tim O'Reilly. He tells me that Simon has influenced everything in his life.

Sitting in O'Reilly's kitchen the next morning, I'm eager to discuss it. But no conversation with O'Reilly travels in a straight line. First we talk about music—one of his daughters is in an indie band in Vermont—and we wind up streaming tunes from my PowerBook to his AirPort Express–enabled stereo. Then we sit down at his computer, and O'Reilly shows me a custom software visualization he uses to track sales of books and inform his radar on what topics are gaining steam in the infosphere. After that, O'Reilly raises another subject close to his heart: scones. With a little remixing, he mastered a family recipe. He pulls some from the oven. Fantastic.

As I take stock of the cluttered living area with the flooring pulled up (awaiting a renovation that looks like it's not happening in much of a hurry), the music thrumming, and the computer fractalizing complex visualizations of book sales, I get a good sense of who this guy is. But you can't get the whole picture of the guru of participation until you hear about *his* guru.

George Simon, born in Germany in the mid-1920s, was raised in New York City and wound up in California, where, as O'Reilly says matter of factly, "he sold toilet paper." After studying Zen, semantics, and yoga, Simon came up with the idea of building a language for consciousness. He was also a scoutmaster, and in the mid-1960s the boys in his Explorer troop became his students as well. Among them was O'Reilly's older brother Sean, and he brought in Tim.

O'Reilly was drawn to Simon's eclectic philosophy. Bright and curious, Tim grew up in a big family in San Francisco—three brothers and three sisters—and he learned to speak up over a din. But he often found himself at odds with his strict Catholic father. (When young O'Reilly checked out Frank Herbert's *Dune* from the library, his father said it was sinful for such a large book to be written about science fiction.) Although he didn't buy into Catholicism, O'Reilly absorbed its morality. In high school, an aptitude test even earmarked him for the priesthood. But Tim found himself answering a different calling. Even after the family moved to Virginia, the boys would go back to Simon on their school breaks. Instead of making s'mores around a campfire, they would ponder the weirdly mathematical consciousness theories of their leader. A typical passage from O'Reilly's 1976 notes on Simon's journals:

> In the end of Notebook 22, George found that the UZ led on to a C quadrant for HS4. He eventually came to feel that the Z16 covered HS1–3, with a normal HS1 having C and some D, and HS2 having C,D,B and an HS3 having ABCD plus a sense of God.

"When I look back on what Simon wrote, it's such total cultish, wacky stuff," O'Reilly admits. "But the *essence* of what he had to teach was so right on—the core of how I relate to life came from him." It's hard to extract what that essence is, but reading the notebook the previous night, I saw that George Simon had written: "To know where the path leads means to be able to go there." Underneath which young O'Reilly had scrawled, "Absolutely revolutionary."

"The stuff I did with George," he says, "relates to how I think and operate, the big patterns I see." That includes the pattern O'Reilly talks most about today: the collective nature of the Internet. He believes Simon's vision of a global consciousness is fulfilled by today's Web.

Simon's legacy in O'Reilly's life has a more personal component as well. In 1972, when O'Reilly was 18, his mentor none too subtly connected him with another follower, an elementary-school teacher seven years older. This was Christina.

Simon's sudden death in 1973 was a terrible blow to O'Reilly, then a freshman at Harvard. Christina was teaching on the West Coast, and his existence in Cambridge was solitary, dominated by reading. After graduating in three years, he married Christina—causing a serious rift with his father, who didn't approve of a marriage outside the faith. Though O'Reilly was heir apparent to the small but devoted community of Simonites, "I realized I didn't want to make my living being a spiritual teacher," he says.

O'Reilly got a National Endowment for the Arts grant to translate Zen-like fables of the Greek philosophers. Later he wrote a well-received but not lucrative biography of Frank Herbert. By then, he and Christina were living in the Boston area. She wanted a family. He had the knowledge imparted from his guru, his Harvard degree, and lots of stories about Spartans, but none of that would buy a ride on the MTA. Somehow he had to make a living. How could he do that in a way that had meaning? What would be the yes beneath the crushing no?

In 1977, O'Reilly met Peter Brajer, a Hungarian engineer who was taking a class from Christina on nonverbal communication. Brajer, seeking computer-consulting jobs, asked O'Reilly to assist him with his résumé, and the makeover helped Brajer land an offer from Digital Equipment Corporation to write an equipment manual. He proposed that O'Reilly help him, and though O'Reilly had never even seen a computer, they completed the project and then went into business. "I actually feel like I did some of my best work as a technical writer in those early years, because I didn't know anything," he says. "I would just have to read stuff until I saw the patterns."

By 1983, O'Reilly had learned enough about computers to start his own business. He set up shop in a converted barn in Newton, Massachusetts, with about a dozen people, all working in a chaotic open room. "The company then was a loose confederation of people who knew Tim," says Dale Dougherty, who fell into the circle in 1984 and is now O'Reilly's most trusted associate and a 15 percent partner in the business.

What happened in that room was a small revolution in technical writing. The O'Reilly approach was to figure out what a system did and plainly describe how you could work around problems you encountered. "The house style was colloquial—simple and straightforward," Dougherty says. "And the other thing was to tell the whole story, not just what's easy to say."

In 1988, O'Reilly and Associates was producing a two-volume guide to the programming libraries of the X-Windows system; while they were in the process of showing it to vendors for licensing, people kept asking if they could buy single copies. MIT was about to host a conference on the system, and O'Reilly figured he'd give it a shot. "We went to a local copy shop that night and produced around 300 manuals," he recalls. "Without any authorization, we set up a table in the lobby, with a sign saying copies of an Xlib manual would be available at 4:30. By 4 p.m., there was this line of 150 people. They were literally throwing money at us or sailing their credit cards over other people's heads. That was when we went, 'Publishing could be a really big business.'"

As O'Reilly shifted from contract work to the book business, he realized that his guides would have to look appealing in a bookstore. The first designer he hired offered a cover with bright colors and geometric patterns—too slick. One of his writers asked her next-door neighbor, a graphic designer named Edie Freedman, to help out. Freedman thought the baffling names of programming tools—awk and sed and Perl—sounded like weird animals. She created covers that incorporated woodcut drawings of creatures from the Dover copyright-free archive.

The tarsier, with its huge eyes, long bony digits, and head that can swivel 180 degrees, became O'Reilly's unofficial mascot. Freedman chose it for *Learning the vi Editor* because O'Reilly thought the tarsier "looked like somebody who had been a text editor for too long." Some choices would become legendary, like the hideous Surinam toad on *Windows Annoyances.*

The books took off, and soon O'Reilly's firm had dozens of guides and a cult following. In 1989, he added a certain mystique to the company by moving it to Sebastopol—not exactly a publishing nexus. The business thrived, partly because of O'Reilly's naïveté. Rivals were selling books to stores at the textbook discount of 20 percent off list price. Not knowing any better, O'Reilly sold his at a "trade discount" of more than 40 percent, a deal that bookstores couldn't pass up. As a result, they were willing to take a chance on more technical tomes they had previously passed on.

But O'Reilly's radar was responsible for his biggest coup. He sensed that the Internet, then still a province of the government and universities, was the platform of the future. Yet there was no documentation for those willing to venture into cyberspace. O'Reilly found his author in Ed Krol, an unknown gearhead who had written an online document titled "A Hitchhiker's Guide to the Internet." Krol took more than a year to expand his short report into a book. Because it had started as a free digital pamphlet, O'Reilly arranged for the second, lengthier part of the book—a catalog of what's on the Net—to be available free online.

The Whole Internet User's Guide and Catalog (1992) became a category-busting best-seller, establishing itself as "a 250,000-copies-a-year thing," O'Reilly says, at least until it became outdated in the mid-1990s. He saw the book not just as a tent pole for his business but as a chance to awaken the world to the Internet. He went on a press tour. He sent a copy to every member of Congress and was invited to meet with House aides. Before addressing a huge group of them, he was taken aside by the House IT department. "I go into this little room, and it's like *Three Days of the Condor,*" O'Reilly recalls. "This old guy in a three-piece suit and a cane says, 'We don't want you to get the aides too excited about the Internet, because we're not going to give it to them.' So I went out and got them excited anyway."

O'Reilly was learning how to become a successful businessman without suppressing his open nature. Borders exec Carla Bayha recalls meeting him for the first time at a computer-book-publisher's conference. Seated next to her at lunch, O'Reilly excitedly began telling her all the great ideas he had for next year's list. "I had to shush him up," she says. "All his competitors were sitting at the table!"

At first, propelled by what he calls "daddy-juice," O'Reilly had been trying to make a living for his wife and two young daughters, Arwen and Meara. But as he got deeper into it, he realized that even the jargonesque domain of computer books could be a pathway to something more—that unbroken yes he first learned from George Simon. "People don't care about books," O'Reilly says. "They care about ideas." And no one would have more ideas than O'Reilly about the Internet age that was rapidly approaching.

In the early 1990s, O'Reilly and Associates found itself on top of a priceless claim in the emerging Internet gold rush—as not only a guide to cyberspace but an innovator in creating Net destinations. To O'Reilly, the effort was as much about pushing the idea of the Internet—and the changes it could bring to society—as it was about turning a buck. That was perhaps why, despite being among the first to realize there would be Internet commerce, O'Reilly never rose to the top of its food chain.

But he did claim some serious firsts. When someone suggested he start a magazine about the Internet, O'Reilly asked why not start a magazine on the Net? The company's catalog morphed into what was arguably the first portal, the Global Network Navigator (GNN). "We were the first people to do advertising on the Web," he says. "I actually saw in 1993 that the ad could be the content, the destination." In a related venture, O'Reilly offered a pioneering off-the-shelf kit called Internet in a Box to help novices make the jump into cyberspace.

GNN also collaborated on a directory of the Web created by two Stanford students, Jerry Yang and David Filo.

At one point, they asked if GNN was interested in buying the venture—Yahoo!—outright. "I think they wanted a million dollars, and we didn't have a million dollars," O'Reilly says. (Yahoo! doesn't deny the discussion but says it didn't set a price.)

Nor did he have the resources to grow GNN once the media giants jumped in, and so in 1995 he sold it to America Online for $11 million, mostly stock and options. "If we'd held on until the peak, that stock would have been worth a billion dollars," he says.

Still, he has pocketed some nice change over the years by investing in companies that show up on his radar. (As an early funder of Blogger, which Google bought, he wound up with a chunk of pre-IPO stock in the search giant; more recently he's invested in the bookmark-sharing service del.icio.us.) But he's nowhere near as loaded as some others who did well during the boom. "We never had a megascore," he admits. "We could have gone public if we wanted to, but why cash out and get all those headaches?"

O'Reilly's company was hit hard by the crash. In the late 1990s, its expanding publishing schedule required a move from cramped quarters in the center of Sebastopol to a brand-new complex down the road. It took six years to complete the project, thanks to a lawsuit filed against the city by locals who saw O'Reilly as more of a modern-day robber baron than a scone-baking publisher. By the time the new campus opened in 2001, the bubble had popped, and book sales tanked. O'Reilly had to fire some 70 people, about a quarter of his staff. "It was the worst I've ever seen him," Christina says. "Service is a big part of who he is—a big goal is to give people jobs." O'Reilly admits that if he'd gone public, "I could have given them comfortable retirements."

The buildings still have almost 30 vacant rooms, as if there's another tenant that hasn't moved in yet, and annual sales are still $15 million lower than they were during the boom. But the titles continue to pour onto the shelves, revealing which languages and apps are au courant in nerdville.

They serve a range of readers, from the gronks who gobble up the code books to the vanilla users consulting the Missing Manual series or general readers who want to know about privacy issues or the history of the Macintosh. O'Reilly also has an online service called SafariU that lets schools and businesses remix their own texts from O'Reilly publications.

The most successful recent venture is a wonderfully retro idea: *Make,* a quarterly print magazine in the spirit of Boy Scout DIY projects. The first issue, published in February 2005, had articles on doing aerial photography with kites, making your own videocam stabilizer, and building a machine to read the magnetic stripes on credit cards. O'Reilly believes that the urge to hack stuff is "more common than we thought." And it dovetails perfectly with the participation-based Internet he extols. The magazine has already exceeded his goal of 30,000 subscribers.

It's one more confirmation that this is O'Reilly's time, and not just in his professional life. He has always been rooted in his family relationships, and the years that his father refused to allow O'Reilly into his home because of his marriage and departure from the church pained him immeasurably. Ultimately, they reconciled, and at his father's deathbed, O'Reilly tried to explain to the old man that his maverick life—and especially his business—was really about the concepts of goodness he'd learned at home.

His father, who at that point could not speak and had to write his comments on a slate, scrawled what O'Reilly saw as the apology he had waited decades to hear: "I only wanted you to be with us in paradise." Now O'Reilly and his siblings jointly own a castle in Killarney, Ireland, where their father is buried. The seven O'Reilly kids of Tim's generation, and their 39 children, all work on it together. "The idea of a project that brings people together was really what was interesting about it," O'Reilly says. Like castle, like Internet.

Satisfaction also comes from O'Reilly's continuing connection to his other late father, the toilet paper salesman–cum–guru who opened up mental vistas that only recently have reached full fruition in his student—now that the idea of collective consciousness is becoming manifest in the Internet. "The work with George was about the future and the potential of what it is to be human," O'Reilly says. "But here we are. The Internet today is so much an echo of what we were talking about at [New Age headquarters] Esalen in the '70s—except we didn't know it would be technology mediated."

Could it be that the Internet—or what O'Reilly calls Web 2.0—is really the successor to the human potential movement? If so, the new Esalen is his increasingly fabled Foo Camp, where 200 or so people—a gamut ranging from Amazon.com's Jeff Bezos to BitTorrent's Bram Cohen and random wizards doing voice over Internet protocol (VoIP) hacks—are invited each year to the underpopulated Sebastopol campus to crash in empty offices or pitch tents in the backyard.

At the Foo Camp in August, O'Reilly opened by asking participants to identify their passions with only three words. After the introductions came a mad rush to a giant poster board to fill up the empty squares with an instant, self-organized agenda. Sessions ranged from "Beyond Python" to "Tele-operated Humanoid Robotics."

Tim O'Reilly's three words? *Harnessing collective intelligence.*

Adam L. Penenberg

The Right Price for Digital Music

Why 99 Cents per song is too much, and too little

In the early 1900s, jazz musicians refused to record phonograph records because they feared rivals would cop their best licks. We can laugh at their shortsightedness, but it's reminiscent of today's music industry, which is so afraid of piracy it still hasn't figured out how to incorporate digital downloads into a sustainable business model. Each year record companies ship about 800 million compact discs—nearly 10 billion songs. That sounds like a lot until you compare it to the 13 billion songs that were available (according to download tracker BigChampagne) for free on peer-to-peer networks in 2004.

The one bright spot for the industry has been Apple's iTunes store, which has sold 600 million songs since 2003, accounting for 80 percent of legal downloads in the United States. Piracy is clearly here to stay, but as iTunes has shown, the record companies' best strategy is to provide an easy-to-use service that offers music downloads at a fair price. But what price is "fair"? Apple says it is 99¢ a song. Of this, Apple gets a sliver—4¢—while the music publishers snag 8¢ and the record companies pocket most of the rest. Even though record companies earn more per track from downloads than CD sales, industry execs have been pushing for more. One option is a tiered pricing model, with the most popular tunes selling for as much as $3. After all, the music honchos reason, people pay up to $3 for cell-phone ring tones, mere snippets of songs.

Steve Jobs, who has been willing to take a few pennies per download so long as he sells bushels of iPods, calls tiered pricing "greedy." That view is shared by millions of consumers who believe the record companies have been gouging them for years. From the buyer's perspective, however, Apple's 99¢-for-everything model isn't perfect. Isn't 99¢ too much to pay for music that appeals to just a few people?

What we need is a system that will continue to pack the corporate coffers yet be fair to music lovers. The solution: a real-time commodities market that combines aspects of Apple's iTunes, Nasdaq, the Chicago Mercantile Exchange, Priceline, and eBay.

Here's how it would work: Songs would be priced strictly on demand. The more people who download the latest Eminem single, the higher the price will go. The same is true in reverse—the fewer people who buy a song, the lower the price goes. Music prices would oscillate like stocks on Nasdaq, with the current cost pegged to up-to-the-second changes in the number of downloads. In essence, this is a pure free-market solution—the market alone would determine price.

Since millions of tunes sit on servers waiting to be downloaded, the vast majority of them quite obscure, sellers would benefit because it would create increased demand for music that would otherwise sit unpurchased. If a single climbed to $5, consumers couldn't complain that it costs too much, since they would be the ones driving up the price. And enthusiasts of low-selling genres would rejoice, since songs with limited appeal—John Coltrane Quartet pieces from the early 1960s, for example—would be priced far below 99¢.

The technology for such a real-time music market already exists. The stock exchanges keep track of hundreds of millions of transactions every day and calculate each stock to the quarter-penny in real time. Banks are able to do the same with hundreds of millions of ATM withdrawals. A music market would actually be much simpler. When a trader on the Chicago Mercantile Exchange buys soybean futures, he or she has to take into account weather, crop yields, supplies in other parts of the world, and the overall economy. On the Digital Music Exchange, there is only one input: demand.

The interface could look something like Apple's iTunes, where users search for songs they want. One important addition would be a ticker that calculates the number of times a track has been downloaded. Click on the icon to see how much it costs right now. Click again and you freeze the price—we'll give you something like 90 seconds to make up your mind—and make the purchase. If you buy a track for $1, that doesn't necessarily mean the price goes up for the next person. Just like on the stock market, it might take a lot of transactions to move the market. Another potential feature, stolen from Priceline: If you tell the system how much you're willing to pay for the new 50 Cent single—say, less than 50¢—it could send you an e-mail alert when the market is willing to meet your price.

This is all really just a corollary to Chris Anderson's Long Tail theory. In the material world, stores sell goods that generate a satisfactory return on the space they eat up. According to Anderson, the editor in chief of *Wired,* your run-of-the-mill record store has to sell at least two copies of a CD per year to compensate for the half inch of space it takes up on the shelf. But in the digital realm, there is no shelf space. Infinite amounts of product are available. Instead of a hit-driven culture, we experience what a friend of mine calls "an embarrassment of niches." A record company doesn't have to depend on one album to rack up sales of 5 million. They can make the same money selling 500 copies of 10,000 different titles or, for that matter, 5 copies of 1 million titles.

Of course, there are modest fixed costs associated with this pricing model: bandwidth, servers, office space, electricity, and the salaries of people who maintain the business. That means there would have to be a price floor, perhaps 25¢ a song. But each obscure indie rock or klezmer song that gets sold for a quarter is almost pure profit, and the bargain-basement price would induce people to download even more tunes.

The big wild card here is the impact of illegal file sharing. David Blackburn, a doctoral student at Harvard, has argued that peer-to-peer systems increase demand for less popular recordings but dampen sales of hits. If that's the case, charging extra for top sellers might just push legal downloaders back into the outlaw world of peer-to-peer file trading. If that happens, perhaps the record companies will start offering free digital downloads of top-100 hits (with ads embedded inside, of course), while charging whatever the market will bear for the rest. A Digital Music Exchange may not be a perfect solution, but who would you prefer to set the price of music: consumers or record executives?

Dan Ferber

Will Artificial Muscle Make You Stronger?

The world's first human-robot arm-wrestling match shows off the potential of a new material that someday could power machines—and even human limbs and organs.

In the annals of organized arm wrestling, there had never been a match like this. Ever since 1952, when the first official arm-wrestling competition took place at Gilardi's Saloon in Petaluma, California, contestants have generally been large men with unusually muscular forearms. But on this Monday afternoon, the TV cameras focus on a slim 17-year-old girl. Panna Felsen's very participation is odd for a sport that still counts events like the Bad-to-the-Bone Armwrestling Championship in its lineup (and the fact is, she trained harder for her high school's Science Olympiad than for today's event). But even more unusual is her opponent. As match time approaches and the crowd grows quiet in the cavernous hotel ballroom, Felsen glances shyly at the phalanx of journalists surrounding her. Then she walks to the padded regulation arm-wrestling table, places her elbow down, and grips her opponent—a white plastic robotic arm.

The first-ever human-robot arm-wrestling match, held in March in San Diego, marked a milestone—as the emcee,

Yoseph Bar-Cohen, a physicist at the Jet Propulsion Laboratory in Pasadena, was eager to declare to anyone within earshot. Never mind that Felsen's opponent looked more like a beefed-up bowling pin than a real human arm. What mattered was the way it moved. The arm that Felsen was about to wrestle, and two more waiting in the wings, had no gears, no shafts, no cams, no moving metal parts whatsoever—a fact that distinguished them from windshield wipers, disk drives, prosthetic limbs, and the millions of machines on Earth that create motion using electric motors. The robotic arm clamped to the table across from Felsen was propelled entirely by plastic.

The material driving these arms is a little-known one called electroactive polymer that has an unusual property: When stimulated by electricity or chemicals, it moves. It expands and contracts, curls and waves, pushes and pulls. It's also springy, durable, quick, forceful, and quiet. Since those properties are shared by human skeletal muscles, electroactive polymers have been dubbed "artificial muscle."

Artificial-muscle enthusiasts like Bar-Cohen foresee a vast array of cheap, light, versatile, and powerful actuators—motion-generating devices—for military technology, space vehicles, and medical devices. Roy Kornbluh of SRI International, a pioneering artificial-muscle researcher, predicts that the materials could ultimately replace up to half the planet's 1 billion electric motors. Already engineers are developing artificial-muscle-powered devices, including a knee brace that prevents injuries; tiny pumps to deliver drugs; and robots that wriggle like snakes, fly like birds, or hop like grasshoppers.

But beyond those devices lies an even more ambitious goal: to replace the genuine article. "In this material, we have the closest to real muscles we ever had," Bar-Cohen says. Research has begun on a variety of medical devices that would be implanted in or attached to people's bodies, such as artificial-muscle-powered prosthetics, a pumping device to assist diseased hearts, a urinary sphincter to treat incontinence, and an artificial diaphragm to help people breathe. Further—much further—down the road, scientists talk of plastics that could replace or augment any muscle in the body.

Six years ago, Bar-Cohen issued a challenge: Build a robotic human arm that could beat the strongest arm wrestler on Earth. But by 2003, impatient to see his contest actually happen, he eased the requirements. It would be necessary only to prevail over a high-school student, he announced. The rules: Robot arms "should not perform any irritating acts," such as flashing blinding lights, vibrating, or making annoying noises. And to win, the robot arm would have to be able to rotate back to its starting position after pinning the human. Now, at the annual scientific conference on artificial muscle in San Diego, three teams claim to have made such a device. The upcoming event excites Bar-Cohen no end. "It's incredible that we have it at that level!" he raves.

So at just after 5 p.m., Felsen, a senior at nearby La Costa Canyon High School, faces her first opponent. (Bar-Cohen chose her after learning that she'd started an engineering club at her high school and liked to build simple robots for fun.) Projected on a two-story screen at the front of the ballroom is an image of the robotic arm and Felsen, who at the moment resembles a deer in the headlights. Bar-Cohen quiets the crowd in his thick Israeli accent. "OK, go!" he barks. An arm-wrestling champion who had come to see the match has volunteered to help Felsen with her technique. He crouches by the corner of the table. "Stay close. Up on your toes," he says. The robot doesn't budge. Felsen, straining, manages a self-conscious smile. "Push harder!" her coach urges. Felsen's face grows determined, and she struggles to tip the robot arm over.

Ask a bioengineer about muscle, and you'll hear high praise and a spec sheet full of properties. First and foremost is force. A thigh muscle can generate 36 pounds of force per square inch—enough to snap a pine board. Then there's power, the rate at which the force is applied over distance. As in automobiles, high power leads to tremendous speeds; a typical skeletal muscle produces horsepower that pound for pound is "way more than a car engine," says bioengineer Richard L. Lieber of the University of California at San Diego. Muscles also act as brakes, springs, and shock absorbers, which is why we, unlike your typical robot, can run, jump, and land softly. And finally, as a waggish British biologist once put it to a roomful of engineers, muscle is "good to eat."

Artificial muscle will never rival a good rib eye, but it's on its way to replicating many of muscle's other properties. To generate force on command, a material must first be able to deform, like a rubber band, at the flick of a switch. It must contract or expand far enough to move an object a sufficient distance. And it must be stiff enough to generate sufficient force. An effective arm-wrestling robot has to match the force of human torso muscles while rotating an armlike extension and must have sufficient control to adjust its force as necessary. "It's the ultimate in application," Bar-Cohen says. "If I can do that, I can make something useful." The three arms pitted against Felsen each employ a different type of artificial muscle, so the contest will double as a test of the field's most promising technologies.

Felsen's first opponent was built by Mohsen Shahinpoor at Environmental Robots, a small Albuquerque-based company. It's made of ionic polymer metal composites (IPMCs), which require low voltages but move rather slowly. IPMCs are bendable, and so they can be molded into whatever sort of actuator will be most powerful.

If a bookie were laying odds on this match, the favorite would probably be Felsen's second opponent, built by a team of Swiss government engineers. This arm is propelled by dielectric elastomers, films in which thin carbon-based electrodes sandwich a soft plastic such as silicone or acrylic. Electricity draws the electrodes together, squeezing the plastic, which expands to up to three times its normal area in about half a second. Actuators made of dielectric elastomers exert up to 30 times as much force, gram for gram, as human muscle. But they require several thousand volts of electricity—a bit of a problem if you want to use them near, or in, the body.

Felsen's third opponent is the underdog. A team of undergraduates from Virginia Tech University, working long nights on a tight budget, created a gel fiber that shrinks when acid is added. The students couldn't get anyone to donate the artificial muscle, so they made it themselves. Their creation is slow to get going but contracts a lot, up to 40 percent of its length, and has the additional benefit of requiring no electricity.

All three teams arrived early to prepare. As the hours tick down, the pressure mounts.

At the resort where the artificial-muscle conference is being held, palm trees lean over red brick walkways and relaxed tourists mill around in colorful beach clothes. Inside a low-slung building, it's easy to spot the Swiss team—they're the anxious-looking engineers talking together in low murmurs. Their device sits on the floor of a conference room. It's a black fiberglass-composite box.

Gabor Kovacs, the team's lead engineer, has been building dielectric elastomers for five years. His goal is to develop shape-shifting actuators that could be used, among other things, to reduce wind resistance in blimps. But lab tests on the material only go so far; last year he took up the arm-wrestling challenge. "We wanted to see the possibilities and limits of this technology," he says.

After consulting biomechanics texts, Kovacs and his team, who work for a government lab, decided to simulate the torso muscles an arm wrestler uses by rotating the entire robot (the black box) around an axle (a stand-in for the shoulder joint) while holding the "arm" stiff. To make their actuators, they built a machine that stretches a sheet of silicone and sprays it on both sides with a chemical coating. The machine then wraps this three-layer film, a dielectric elastomer, around a springy steel core. To maximize the actuators' power, the team spent months experimenting with various chemical formulations of the coating.

This afternoon, a day before the match, Kovacs expounds on his masterpiece. It took a year to build, he tells me, and cost $250,000 in Swiss government funds. It possesses 256 actuators, powered by up to 4,000 volts. But before he can finish, two young men appear. They confer in German, glancing unsmilingly at the arm 'bot. As they pick it up and prepare to leave, the only words I understand are "Home Depot."

The designer of the beefed-up bowling pin, Felsen's first opponent, sits across a table from me in the hotel ballroom and grips my hand in an arm-wrestling posture. Mohsen Shahinpoor runs the Artificial Muscle Research Institute at the University of New Mexico and directs research at Environmental Robots. In a technical talk that day, he had shown videos of his devices in action. In one, a human skeleton pedaled a bicycle, powered by strips of artificial muscle.

Shahinpoor tightens his grip and tells me to push. As I start to win, he pushes back forcefully. His robot arm is programmed similarly: to add power when it starts losing. A sensor measures the angle between the arm and the table and increases voltage as needed.

Shahinpoor's IPMCs consist of two metal-foil electrodes sandwiching a wet, Teflon-like plastic soaked with lithium ions. Just 12 volts—the equivalent of a car battery—cause the lithium ions, which are positively charged, to migrate toward the negatively charged foil layer, bulking up that side of the actuator and bending the IPMC. IPMCs are safer than dielectric elastomers because they use low voltages, and they're stiffer, which enables them to exert more force per actuator. But they're more sluggish because they're triggered by bulky ions, not speedy electrons.

Environmental Robots spent $24,000 to develop the arm (more if you count Shahinpoor's time). He tells me that he limited the machine's voltage and force; he's more concerned about demonstrating its potential, and about the human contestant's safety, than about winning. When we finish talking, he heads to the event. Soon Felsen is pushing against his 'bot as a bemused crowd of about 150 looks on. She is using her arm, not her whole body, and she's struggling. "Push harder!" exhorts the arm-wrestling champ. Twenty-six seconds into the match, she pins the robot arm. She flashes a big smile as the audience applauds. Now there are two teams left.

Three clean-cut engineering students from Virginia Tech—Steve Deso, Stephen Ros, and Noah Papas—worked evenings and weekends in the lab for months to build their robotic arm as part of a project required to graduate. "We wasted our entire senior year—no partying,"

Deso says, and the others laugh. They spent three years in class learning about Newtonian mechanics, solid mechanics, and biomechanics and hanging out together in their spare time. "We were looking for something that would apply our skills," Ros says. They spotted an online article about the contest and joked about entering. It seemed too big a project at first, Deso recalls. They e-mailed Bar-Cohen. "He said, 'Just go for it,'" and they did.

The three decided to use an artificial muscle called polyacrylonitrile, a gel imbued with fibers for strength. After burning through their $800 budget, they begged, pleaded, and scrounged parts and help. Unable to afford polyacrylonitrile, they synthesized it, starting with textile fibers donated by the manufacturer, Mitsubishi Rayon. A prosthetics company donated a metal elbow for the arm, and a body shop spray-painted it for no cost in maroon and orange, their school colors.

Papas pulls one of their muscles from a plastic bag to show me. It's about a foot long and three-quarters of an inch thick, brown and moist; it feels like a piece of raw meat and resembles a giant slug. For the match, they'll place it inside Plexiglas and use a windshield-wiper pump to spray acid on it. They'll tie the ends of the muscle to the arm with 50-pound-test fishing line so that it can wrestle. If all goes well, the acid will penetrate the gel, neutralizing charged groups in the plastic and forcing the material to contract. The students know their muscle is strong enough to perform, but they haven't had time to test their final configuration. When the match is over, they're heading to Vegas.

The Swiss engineers clamp their arm to the table. They connect red and black leads from the box to a bank of power sources, and Felsen dons a heavy rubber glove for safety.

Bar-Cohen walks to the table, microphone in hand, and booms, "It doesn't look like an arm, it doesn't feel like an arm, but it is an arm!" On three, the black box emits a low hum and the scent of burning rubber. Taking pointers from the champ, Felsen uses her body this time—and downs the arm in four seconds. The audience laughs. The Swiss team, which hadn't had the chance to fully test their creation, looks on, stone faced. "I thought we couldn't lose," Kessler would later tell me.

Now it's the Virginia Tech team's turn. The students, dressed in crisp-looking maroon polo shirts, clamp their project to the table. "Now it gets exciting," Bar-Cohen booms. He hands the microphone to Deso, who tells the crowd how they're going to activate their device by adding acid. Felsen dons safety goggles and grips the hand. She pushes, and in three seconds the arm flops to the table without even a hint of resistance. A few moments afterward, inside the Plexiglas contraption, the sluglike artificial muscle begins to contract. "Our goal was just to get here," Deso tells me. "That was a huge accomplishment." And they do seem happy. Only Ros voices a regret—that they didn't activate their muscle earlier: "If [the match] had started five seconds later, we could have put up a fight."

The first-ever human-robot arm-wrestling match seemed to mark an ignominious defeat for the robots. A girl who describes herself as "not very strong" had trounced some of the best artificial muscles that engineers have to offer.

But the researchers continue undaunted. Kovacs is back to making dielectric elastomer sheets that curve into complex shapes on command. Such a material could enable aircraft wings to change shape in flight or could emulate the undulating fins of a stingray to propel underwater vehicles. And Shahinpoor's company is developing two medical devices: an adjustable band that would correct nearsightedness by squeezing the eyeball to alter its curvature and length and a device that would help ailing hearts pump blood.

Just before I leave, I spy the Swiss engineers in the exhibit hall. They have taken apart their robot and clamped one of its actuator banks to a table. Two carboys, heavy with gallons of water, hang from the actuator bank. A sign with large red letters reads, "Caution! High voltage!" With an air of mild disappointment, Kovacs explains that the rule requiring the arm to spring back prevented them from demonstrating their machine's full strength. To simulate human muscles, which operate in pairs, one contracting while the other relaxes, the team had aligned their actuators to work in opposition, canceling each other out—except for a small differential, which represented the arm 'bot's force. But now, released from the machine, their actuators could show their true power. One of Kovacs's colleagues flips a switch, and the artificial muscle hoists and lowers the carboys—up and down, up and down.

"We didn't win," Bar-Cohen acknowledges. He adds: "Twenty-six seconds is maybe nothing." Then, his voice rising with excitement, he continues, "But the first flight was 12 seconds. We have to remember that. A hundred years from now, who knows where we could be?"

Richard Waters

Plugged Into It All

From Japanese girls texting their friends to the BlackBerry-armed executive, it's now who you are connected to, not who you know. But does this mean greater freedom or loss of control?

To listen to a group of students at the University of California's Berkeley campus talking about their obsessive communications habits, you would think you had stumbled into a meeting of recovering alcoholics. Rich Brown, a graduate student at Berkeley Haas School of Business, confesses to being forced into drastic action the previous evening when, at 10 p.m., it was time to get down to some serious work. An instant messaging exchange had to be terminated: e-mail, a constant companion, was shut down. "But it's a compulsion," he says. "I had to check my e-mail half an hour later. You have to look at it again."

Laptops are lined up, closed, on the table in front of them; the occasional mobile handset placed alongside like illicit drugs that they have been asked to surrender. These students betray the modern ambivalence of the constantly connected: pride in their technological virtuosity mixed with a self-consciousness about their infatuation that pushes them to joke about their condition.

/Financial Times/

"I think it's horrible," says Christian Oestlien, a fellow student and self-confessed e-mail fanatic. "Once you get in, you can't get out."

Tools such as e-mail and instant messaging may have been around since the dawn of the Internet era, but it has taken a wireless communications revolution to turn them into a constant and inescapable fact of life for a growing part of the population. WiFi networks—a low-cost technology that can beam large chunks of data over short distances using part of the radio spectrum that was previously the preserve of gadgets such as garage door openers and baby monitors—assure the digitally addicted of a permanent and ubiquitous connection to the wider world. At the same time, more versatile mobile phones have turned text messages into the communications tool of choice for teenagers in Asia and Europe, if not yet the United States, while also bringing e-mail to many handsets. For those in the grip of these new networks, life has changed. There's no such thing as solitude any more, no fragment of time that cannot be filled with digital chatter.

Students are less likely to work in the library, says another Haas student, Sung Hu Kim—it's one of the few places on campus where the WiFi signal is weak. Work happens anywhere there is wireless access and a comfortable place to sit: on the grass outside the faculty buildings or slouched in the student lounge. In the lecture halls, meanwhile, laptops are kept open—unless a professor objects—and instant messaging and e-mail services are left connected, to be checked fleetingly and often.

Even these self-confessed communication junkies may not be ready for the full-time commitment that avid exponents of texting often display. Richie Teo, a graduate student from the Philippines, says he's surprised by the lack of texting among his American counterparts: back home, with his techno-savvy friends, he was accustomed to getting and sending dozens of messages a day.

"If I've been to sleep and don't have at least four messages when I wake up, I feel no one loves me," he says. The aghast looks of his peers suggest that these are depths of communications addiction that even they have yet to plumb.

For students at campuses across the country, being permanently connected means rethinking where and when to do things. On the other side of the country, the Massachusetts Institute of Technology near Boston has just plugged in the last of the 2,800 electronic nodes that will bathe its 168-acre campus in high-speed wireless internet access. The Steam Cafe (sample menu item: Malaysian fish curry with organic short-grain brown rice) is the new place to hang out while working, says Carlo Ratti, a research scientist at MIT.

"This used to be only used 3 hours a day: now it's 24 hours a day," he says. Classrooms and libraries are emptier as a result.

When you're hooked to this network, the possibilities multiply. MIT plans to let its students continually broadcast their whereabouts to anyone in their personal social network, says Ratti. Overlay that information on a map of the campus or town, and you could keep track of your family or friends all the time. What happens when you free students of the need to sit in lectures—or office workers of the need to be at their desks—and place them instead in a free-flowing, virtual community? What are the implications for the way people work, the way social and political life are organized, and the way cities are run? "Real estate value will be based not on the square footage, but on usage," predicts Ratti. "We won't be working from home—we'll be working from anywhere."

WiFi networks are starting to creep over civic space as well. Cities from Philadelphia to Seoul are planning city-wide networks that would give low-cost or free Internet access to residents. These short-range networks are part of an invisible electromagnetic mesh that is settling over everything. Together with the new souped-up networks being launched by cell phone companies, they are part of a rush to turn the radio spectrum into an all-enveloping blanket of digital communications and new wireless media.

Of course, predicting technology revolutions is foolhardy. They never pan out the way that the visionaries predict and seldom yield the sort of instant new markets of which business planners dream. Telecommuting has been predicted since the beginning of the PC era and seems as quaint these days as the personal jetpacks that we were all meant to be wearing sometime around the end of the 20th century. Ubiquitous mobile networking may yet prove to make just as little impact on our daily habits.

Yet it is hard to deny the extent to which mobile phone communications have already crept into many, if not most, corners of our lives: children texting from the bus stop; suburban streets clogged with housewives on the phone while at the wheel (at least in countries where it is still legal); executives bowed, fetishistically, over their BlackBerries. In equal parts liberating and intrusive, the mobile phone has changed the way many people relate to their work or to their friends and loved ones. It seems a fair bet that its next incarnation will have a much deeper and wider impact.

Most technologies, as they reach a bigger sphere of people, become less widely used, says Glenn Woroch, an economist at the University of California at Berkeley. That has not been the case with mobile phones. The amount of time the average person spends with his or her mobile is going up, he says, even as the network expands.

This is the beginning. The mobile phone is already morphing into an all-purpose messaging device, capable of catching and transmitting many of the minutiae of daily life, from the short snippets of text messages to impromptu photos. Laptops on campuses such as Berkeley and MIT are becoming windows into digital media.

"This is like watching the beginnings of the World Wide Web," says Dick Lampman, director of Hewlett-Packard's research labs. Trying to predict exactly how this personal communications revolution is going to change your life is likely to lead to the same kind of hyperbole—and mistakes—that characterized the early dot-com days, he says, but "you can see the early pieces of it, joined up, in the mobile phone world."

The virtual world is no longer behind a TV screen or on the PC: It's with you all the time. The persistent chatter and, increasingly, the songs or TV shows being streamed over these networks are starting to seep into many aspects of everyday life.

To understand just how deeply mobile communications may eventually affect your life, it helps to consider the habits of Japanese schoolgirls.

What Kenichi Fujimoto, a researcher at Keio University in Japan, calls the "schoolgirl pager revolution" remains one of the most revealing technology events of recent years. Simple numeric pages, designed for business use, were taken up in the early 1990s by teenage girls, who used them to send coded messages to each other. That became one of the models for the short text messaging that now seems to define youth culture.

It was a seminal moment for the technology industry, a sign that the forces of technological innovation had been turned on their head. New technologies had always been created for business use first, on the assumption that employers would be prepared to pay for gadgets that made their workers more productive. That was how the first bricklike mobile phones got their start. Now, though, it is consumers—often teenagers—who are the early adopters of many new technologies. The rest of us follow their lead.

That suggests that the place to look for signs of what we'll all eventually be doing with our mobiles is among young people on the streets of Tokyo, Seoul, or Helsinki.

The evidence is hopeful. Some of the early media coverage of mobile communications in Japan, as elsewhere in the world, pointed to a futuristic dystopia, a place where ubiquitous personal communications would cause the disintegration of social norms. Some Japanese girls were found, for instance, to be using their mobiles to engage in prostitution with middle-aged men. Newspapers were full of stories of mobile dating services that could connect two people who happened to be walking down the same street at the same time.

In a stable, paternalistic society, the power that the mobile gave to the young amounted almost to a social revolution, according to Fujimoto. Ko-gyaru, as the new band of brash, fake-tanned, and dyed-blonde schoolgirls were called, represented a direct challenge to the fathers who held social power. Just talking loudly on a mobile phone on a bus or a train amounted to a rebellion, writes Fujimoto. The mobile-wielding, miniskirted Japanese schoolgirl became a symbol of the technology's power to disrupt social norms.

Other early signs of how the mobile might change behavior added to this sense of social norms unraveling. The ability of groups of people, operating independently, to coordinate their actions appeared to create the chance for political action that welled up from below, outside normal institutional bounds. "Smart mobs" or "flash mobs," the name given to these ad hoc meetings, were at once empowering and scary.

And even if you discounted revolutionary visions like

these, there were more prosaic reasons to fear mobile phones. They are, after all, a distraction, just another thing to divert the attention of multitasking youth or steal away what little remains of our free time.

In fact, the evidence of how most Japanese teenagers use their mobiles suggests that pervasive communications are strengthening social bonds, not breaking them down.

Mizuko Ito, an associate professor at Keio University, has applied the techniques of anthropological research to the study of the use of technology among Japanese youth and concludes that mobile networks are creating a new form of "full-time intimacy." Most people use their phones to stay in close contact with between three and five loved ones or friends, she says. Sociological literature, which has a habit of sprouting important-sounding titles for any new phenomenon, has invented a name for it: "tele-cocooning." The very nature of much of the mobile texting that goes on suggests that its real intention is to act as social glue, maintaining intimate connections between people: as a glance at any teenager's stream of text messages shows, it is seldom to communicate meaningful information.

Text messages are used to fill the dead time, a form of small talk that fits into the gaps in people's lives. Much of the communication is fragmentary and inconsequential. It operates, at a microlevel, as a constant stream of pointless babble.

This persistent, low-level form of contact is really all about maintaining a sense of constant "presence" with people who are elsewhere, says Ito. The virtual world created by the mobile is a shared social space, something always with you: the point is to be always on and always connected, even if right now you have nothing much to say. That suggests that the seemingly pointless, reflexive text messages that pass back and forth are primarily a way to reinforce a social bond and a sense of presence.

Ito compares this to the conscious and unconscious body language that passes between people who are in the same physical space. The seemingly pointless short text message, she writes, is "a sigh or smile or glance, a way of entering somebody's virtual peripheral vision." It may not lead to a conversation, but it is a way of maintaining contact. Users of instant messaging on PCs are already familiar with a version of this phenomenon: it is the sense of presence that comes from the buddy list.

Armed with this persistent connection to a small group of people, Japanese teenagers have learned how to move smoothly between the real and virtual worlds, says Ito, who coedited the book *Personal, Portable, Pedestrian: Mobile Phones in Japanese Life.* They draw their mobile relationships into the foreground when they have time to kill or something to communicate and then push them into the background again when something more immediate claims their attention.

Of course, there are new social obligations that come with all of this. There is an expectation that intimates should be "available for communication unless they are sleeping or working," she says. And even working is not always an excuse—certainly not classwork, given the way many Japanese children keep their phones on their desks at school. Japanese teenagers say that messages have to be returned immediately, or at least within 30 minutes, or a social convention has been violated. Forgetting to take your mobile with you or letting the battery die are considered among the greatest of social misdemeanors.

Texting has emerged as a way to make mobile communications more constant and pervasive, while also reducing the disruption to the experience of "real" life that telecocooning implies. Sending a text message, for instance, means you no longer have to annoy the person sitting next to you on the bus by talking on your phone. In Japan, texting has become the socially acceptable way to stay in contact while on public transport.

It has also become a way of easing the transition from the real to the virtual world. Before dialing someone's telephone number or in the run-up to a face-to-face meeting, a stream of text messages lays the ground. "You don't make voice calls without checking availability first," says Ito. "The ringing telephone is quite a rude thing." With proper texting etiquette, the phone only rings when you want it to, and face-to-face meetings are choreographed by an elaborate ritual of advance messages.

These persistent, mobile-powered social networks fit into a view of modern life that has been gaining acceptance in academic circles. It holds that, contrary to what you may have thought, we are not living in the Information Age: we are living in the Networked Age.

Expressed most fully by Manuel Castells, a Spanish sociologist, the network-centric view of life suggests that we each exist as a "node," or an element, in many intersecting networks—of family, work, and friends. According to this view, humans are defined by the networks of which they are a part. It is no longer what you know that counts: it is who, or what, you connect to. Thanks to mobile communications, we can all soon expect to be connected permanently.

Teenagers may be happy to live in this permanently connected world, but what about those who remember life without the repetitive sound of novelty ring tones?

A decade after mobile phones became commonplace, attitudes are still sharply divided. "The absence of constant connectivity and multitasking is a deprivation for the young," says Lampman, at Hewlett-Packard. For many others, the ability to unplug can seem an essential precondition for sanity in the modern world. For the average office worker, the same sort of social pressures that have been at work in shaping Japanese youth culture are also starting to influence working practices and impinge on home life. The power to pick up—and respond to—e-mail and messages from anywhere is blurring the lines between the office and the rest of life.

For the average executive in Silicon Valley, this has lengthened the workday, says Steve Barley, a professor at Stanford University who specializes in the organization of work and the impact of technology. He thinks this could be "the equivalent of three and a half extra weeks a year just communicating outside work: that's more vacation than most people get" (at least in the United States). Yet by his estimate, this has done little to make workers more productive.

The main reason for all these extra unproductive hours "seems to be a fear of what will happen if you don't check your e-mail before work and in the evening," he says.

Paranoia is rife. Among the connected white-collar classes, it is now no longer acceptable to leave the mobile—or the BlackBerry or Treo—behind or let the battery die.

Permanent access to multiple forms of communications is also producing an addiction to multitasking among members of the professional classes that is inevitably eating into the quality of work, according to Barley. Taking part in a conference call, for instance, is now an excuse for the exercise of minimal attention.

"The game is to pay just enough attention on the telephone so that you can respond when your name is mentioned and keep track of what is going on," says Barley. "This seems to be fairly widespread among the professional managerial class. These things must make meetings less valuable."

Yet these permanently connected executives end up feeling more harassed than ever. Managers who advertise their mobile phone numbers on their business cards and leave their phones turned on all the time are the ones who are most likely to feel overloaded by work, says Barley.

The mobile's facility for filling in the empty hours—and chipping away at the productive busy ones—may be only just beginning. Mobile phones are turning into ubiquitous media devices. Technological advances on a wide range of fronts—faster wireless networks, longer battery life, more powerful processors and memory chips—are conspiring to turn the small voice communicator in your pocket or handbag into a high-powered computer, capable of processing, storing, and displaying all types of media. That may make it the next iPod, a screen for catching up on TV shows you missed last night, or a way to tap into all the photo-sharing Web sites and personal blogs that your nearest and dearest use to chronicle their lives.

This permanent exposure to digital media and communications could really start to change the way you experience your life. And as with the arrival of that last great intruder on personal time—the television—it certainly has its detractors.

In his 2002 book *Smart Mobs: The Next Social Revolution*—one of the best explorations of the impact of mobile communications—Howard Rheingold quotes Leopoldina Fortunati of the University of Trieste on the insidious way that texting has started to consume the idle minutes. "Time is socially perceived as something that must be filled up to the very last folds," laments Fortunati. This modern obsession threatens to eliminate "the positive aspects of lost time" that "could also fill up with reflection, possible adventures, observing events, reducing the uniformity of your existence, and so on."

For workers—and their employers—that lost time has a harder economic edge to it.

"Access to a worker—even a colleague—is a scarce resource," says Glenn Woroch at Berkeley. "Every scarce resource should be priced, either explicitly or implicitly. The fact that you keep your cell phone on, and check your e-mail and your instant messages at your desk, is setting too low a price for this scarce resource." The trouble is, there is no market mechanism for rationing out your time. The mobile phone is an all-or-nothing thing. "To the extent you don't want to be forced out of the network, you're kind of compelled to keep it on and keep it with you all the time," concedes Woroch.

How, then, to limit all the intrusions and regain control of your life?

Some people are developing their own ways to shut off, compartmentalizing their lives into work and social time by consciously creating separate spheres of communications. Some have multiple cell phones for different parts of their lives, just as they have multiple e-mail accounts, says Barley. Research, he adds, shows that managers are by far the worst at segregating their lives in this way and the most likely to allow unproductive work intrusions on the rest of their lives.

Even the creators of this cornucopia of digital goodies concede that it is all getting too much for most users, but, ever optimistic, the technocrats say that the tools of technology will eventually sort this out.

Ray Ozzie, a chief technology officer at Microsoft and one of the pioneers of the use of e-mail and other so-called collaboration technology to organize work life, says that the next five years will see a drive to give workers more control over their communications. Otherwise, tools that are meant to improve the productivity of the average worker—and the quality of life in general—could end up having the opposite effect.

The current Big Idea in technology circles for handing back control is to somehow embed your personal preferences in the technology in a way that makes it respond to how you want to live your life. The phone, for instance, will be smart enough to know when you can be interrupted and when to leave you alone. It will know, on any particular day, whether to put through a call from your mother immediately or whether to send her straight to voice mail.

By learning from the preferences of your close networks of family and friends, it will also have an idea of the sorts of things that are likely to interest you. In the future, we will all be part of "self-organizing peer groups that provide ways to filter things," says Lampman. Only communications or media that have a place in this more closely defined social realm will be able to find their way onto your personal communicator. "People want more control: you should be able to build your own profile and use that to qualify the things that come to you," says Lampman.

One day, this could represent a nirvana for mobile communications. It would be a golden age for personal freedom and choice, the apotheosis of what sociologist Barry Wellman calls "networked individualism"—the power to plug into any number of networks without being subsumed into intrusive and suffocating social groups.

For now, though, there is a far more immediate answer to that insistently ringing mobile in your pocket: just switch it off.

Jim Rossignol

Sex, Fame, and PC Baangs

How the Orient plays host to PC gaming's strangest culture

Seoul, South Korea. To a fanfare of Asian nu-metal and the sound of a thousand screaming fans, a young Korean man enters a dazzling arena. Like an American wrestler at the heart of a glitter-glazed royal rumble, he strides down a ramp toward the stage. Adorned in what appear to be a space suit and a large white cape, he heads out to meet his opponent on the stadium's ziggurat focus. Amid a blaze of flashbulbs and indoor fireworks he climbs the steps and is exalted by the thronging crowd. Only 20 years old, and with no less than half a dozen TV cameras tracking his progress, this bizarre figure seems to be unfazed by his predicament. Diligently he waves to the crowd.

My interpreter, the amiable Mr. Yang, leans forward. "To my brother he is a great hero. My brother can't get enough of this. He has been to see him play many times."

"So this guy has a lot of fans?" I say, knowing the answer but nevertheless incredulous.

"Hundreds of thousands in his fan club," says Yang. "Impossible to track the number of people who watch him play."

Impossible, because the man on the stage is on Korean television almost every day. He is about to sit down and play what is close to becoming Korea's national sport: Starcraft. His name is Lee Yunyeol, or in game [RED]NaDa Terran. He is The Champion. Last year his reported earnings were around $200,000. He plays a seven year-old real-time strategy (RTS) game for fame and fortune, and to many Koreans he is an idol. Every night over half a million Koreans log on to Battlenet and make war in space, many of them with dreams of becoming like Yunyeol. But his skill is almost supernatural. Few people who play all day long will be able to claim a fraction of his split-second timing and pitiless concentration. Yunyeol practices eight hours a day, and his methods and tactics are peerless. Well, almost peerless. In fact there are two or three other players who command similar salaries. They might not hold the crown now, and one of them will probably take it from him soon, but for now at least, Yunyeol is king.

The existence of people like Lee Yunyeol ensures that South Korea is unlike any other gaming culture on Earth. Here the PC is the most important games machine, and major corporations such as Samsung and Fila will pay thousands of dollars to have their logos adorn the best players in the country. This is a culture in which 1 in 20 people has played a massively multiplayer online (MMO) game (it's less than 1 in 70 in the United Kingdom) but where Half-Life 2 barely raised a mention. Regularly playing online sessions of the Blizzard games, such as Diablo II, Starcraft, or Warcraft III, is more common among Koreans than owning a PlayStation. This is a country in which having a subscription to an online game is becoming the rule, rather than the exception. There are five cable channels devoted to games and one of those just to RTS titles like Starcraft. Recorded and edited bouts of top-level Starcraft matches account for viewing figures in the millions, taking up 1 percent of all the TV watched in Korea. There are two weekly newspapers and three four-hundred page monthly glossies that cater just to PC gaming. There are 26,000 gaming cafés in Korea, which make $6 billion a year from tens of thousands of visiting gamers. Seoul is nothing less than a PC gaming hotbed of imagination-defying magnitude.

Intent on uncovering the true story of Korea's peculiar obsession with PC games, I boarded a Boeing bound for that sprawling metropolis and undertook a mission to see it all.

KOREA'S GOT SEOUL

Seoul is at the epicenter of a new East Asian game culture. From Bangkok to Tokyo, millions of Asians are logging on to online games. Their ranks, particularly in the growing economy of China, are rapidly expanding. In South Korea alone, there are nearly 2.5 million people playing online games on a regular basis, and it's the inertia of South Korea's vivacious game culture that is forging the path that other Asian countries are following. As broadband Internet proliferates in Asian countries, so does the South Korean arsenal of online games. Companies like Blizzard Korea, Webzen, NCSoft, and NetMarble are ensuring a future for PC games, and it's a future quite unlike anyone in Europe or the United States would have conceived of. Out here Star Wars games are barely recognized, and while pro games of FIFA occasionally turn up on TV, Halo and Vice City are practically unheard of.

So how did this happen? In this modern world of global marketing, how could one country turn out so at odds with the rest? Well, it has taken some unique economic and political circumstances to make this strange situation a stone-cold reality. South Korea has had a turbulent history and has long endured an intense rivalry with nearby Japan, thanks to the Second World War and previous decades of rampant Japanese imperialism. This rivalry led to decades of trade restrictions that made early generations of Japanese game consoles prohibitively expensive for Korean gamers. If you wanted to play video games in South Korea, then the cheapest way has long been to use a PC. During the 1990s, when gaming first really took off, Koreans were tucking into much the same feast as the rest of the world, only their menu didn't include Sega, Sony, or Nintendo.

But there is another far more significant factor that defined how the Koreans approached gaming. The nationwide focus on the development of cutting-edge technology led their newly democratic government to seize the potential in broadband communications, and, in the late 1990s, they used the then state-owned telecommunications company to install the infrastructure necessary to connect almost every building in Korea to high-speed broadband. The Koreans have taken to the Net with the greatest of ease, with 60 percent of households boasting a broadband connection, compared to just 17 percent in the United Kingdom. It's almost like the Net has been in Korea forever, such is the ubiquity of high-speed connections. Walk past a Burger King in one of Seoul's teeming malls and you'll see people logging on as they munch on a Whopper. It's everywhere, and it's cheap.

This investment was to have a knock-on effect on the growth of small businesses. After Asia's economic downturn in 1997, people were looking for cheap ways to set up a business, and one option was to create an Internet café. Pretty soon every town and city in South Korea were packed with cybercafé start-ups. In the United Kingdom and United States there has been a small but concerted growth in the existence of Internet cafés and gaming centers over the last decade, but in South Korea their rapid expansion was practically a cultural revolution. These gaming cafés, the "PC baangs," soon became the key centers for a youth culture thirsty for social activity and cutting-edge entertainment, and there are now 28,000 of them across the country. For around 50 pence an hour gamers can log on in a baang, play a few games, smoke a few cigarettes, drink some Lineage-branded Coca-Cola, and eye members of the opposite sex. For a culture still steeped in conformity (almost all cars are black, white, or silver, and the middle apartments in a block are the most desirable), games provide a unique outlet for personal expression and a rare chance to be different. PC gaming is not the provenance of a hard-core few, as it is in Europe and the United States; instead it has become the major form of entertainment for young Koreans right across the peninsula. Social, cheap, and available to everyone, online gaming has taken this nation of 45 million people by storm.

KINDS OF LAG

You know that a day is going to be a strange one when, awake in a Seoul hotel room at 3 a.m. (reeling with the worst kind of jet lag), you turn on the TV to witness ex–*PC Gamer* editor James Ashton talking about what a great game character Mario is. "It's just so much fun," says Ashton, without enthusiasm. This chance fragment was simply a random moment in a torrent of repurposed Western TV that turns up on Asian screens, just another node in the international marketing machine. For the most part, however, Korea has little need for the Western (or Japanese) approach to gaming TV. Flip through the five Korean gaming channels and you'll see Starcraft being played for hours at a time. Fast and frenetic, the games are accompanied with the pogoing voices of hysterical-sounding pro-gaming commentators. I was to be informed that these video game anchormen have reached the same sort of status that sports commentators enjoy back home, their distinct characters and ability for smart observations as beloved and revered by the Korean youth as *Match of the Day* pundits are by soccer hordes in the United Kingdom.

Flick to another channel and we plunge headfirst into a stream of advertisements for games that are unseen outside of Korea, all of them online—the cell-shaded Freestyle Basketball, the apocalyptic fantasy role playing game (RPG) Archlord, and a host of indecipherable super-cute games that leap out of the screen with ultrastylized chipmunkesque squeaking. Then it's time for a show in which couples who met while playing Lineage II talk about their love for each other, before doing in-game battle with other couples for fame and prizes. The duels are smartly edited spectacles of pixel magic, and I have almost no idea what is going on.

Korean TV, you should understand, provides a barometer for the rest of culture. Games are so popular among Korea's youth that you often have to be a gamer to be able to socialize—the content of all Korea's game media reflects that fact. TV shows are often designed purely to keep gamers in the know, while gaming magazines and their counterpart Web sites are all about making sure players have the latest tips and walk-throughs, as well as telling gamers how their favorite Korean rock band regularly plays online games. Gaming is so cool that it's practically mandatory, and being good at games can be a great social boon. The hugely popular Lineage II is even being touted as a way for young people to meet partners, with Valentine's Day events held inside and outside the game earlier this year. That's not to say that the only games people play are Starcraft or heavy-going RPGs—for the most part people still play casual games, such as the Mariokart redux Kart Rider, golf games, or one of Korea's many minor Anime-eyed fancies. But however lightweight the game, it is still online, and almost always about playing with other people.

BAANG THE MACHINE

Escaping from the hotel and heading out onto the streets, I leave the business district behind and seek out baangs downtown. I find a few, poster-smothered, smoky, and slightly intimidating. Not in the way that the scary biker pub at the end of my road is intimidating but in that way something familiar made alien is intimidating. It's just a Net café, almost like the ones back in Blighty. There are rows of PCs in a room with a few soda machines and a bored looking girl sitting at the front desk. It should be home territory, but I know I don't belong here. I wonder if it's just because I can't speak the language and most Koreans are hesitant about trying to speak English. Perhaps its because the games are unfamiliar: the English names on game posters seem incongruous above a splash of Korean text.

I wander past a few scattered gamers. Even at 10 a.m. there seem to be some committed souls plugging away at the beasts of Lineage or Mu Online. But that wasn't as incongruous as I'd built it up to be in my head. I knew that Korea gamers were obsessive on a scale we have made legendary, but here they were, casually smoking (in the well-signposted smoking section) or chatting languidly about Starcraft tactics in the gloom beneath a tatty Warcraft poster. These baang customers rarely have to buy games; they simply create an account and then pay for Net time in the baang, with the café manager sorting out the license fees. This means that genuine game retail stores are limited and unusual—

everything gamers need can be downloaded online to play at home or is already installed on a PC baang machine.

This is what is different about Korea: in the United Kingdom you're most likely to encounter other gamers while at the games shop in town, browsing the new Playstation releases, but in Korea you're most likely to encounter them while out playing. That's something that only a tiny fraction of American or European gamers experience, those who go to local area network (LAN) events or a local Net café on a regular basis. It struck me that baangs are no different at all than LAN cafés and gaming centers anywhere in the world; it's just that there are a lot more of them in one place. Thanks to their popularity and the traffic of youth through them, their credibility is much greater. In fact, their place in the cultural consciousness of all of Korean youth is greater. The cool kids wouldn't be seen dead in a UK LAN joint, but in Korea couples regularly go to baangs, and the walls will often be densely flyposted with gig lists or fashion advertisements. Standing there in the Korean baang, it doesn't feel strange at all: It seems like this is just the way it was supposed to be. Indeed, I think back to playing at the gaming center in Bath. Would those guys be in the slightest bit surprised that this is how the Korean youth spends its time? It's the same kind of escapism, the same kind of detached socializing, but on a vast scale. I wondered if perhaps this was a reflection of a deeper national psyche. Korea has always been a shy, inward looking country (the Korean peninsula was once known as the hermit kingdom), and so perhaps their greater obsession with digital entertainment was some manifestation of that humble Korean ego. These people want to be sociable, to have things to see and do, but many of them have turned to games, rather than bars and clubbing, to find that solace.

Further it seems as if more and more Koreans are heading back to their homes to play, now that PCs are cheaper and broadband can easily be installed at home. I sincerely hope that the popularity of baangs isn't simply a blip that will fade with time. I'd hate to think that Korea would be known as the hermit kingdom for anything other than historical reasons.

CARE AND COMMUNITY

Later that afternoon I pay a visit to the customer service desk of online gaming giant NCSoft. The company, profoundly committed to providing a comprehensive service to its hundreds of thousands of customers, is proud to be able to offer face-to-face consultation in its customer care. The help desk for Lineage and Lineage II is a discrete office in the teeming hi-tech district of southern Seoul. The manager comes out to greet me, along with my interpreter and the NCSoft publicity manager. The staff here represent the last word in care for aggrieved players, and the small team deals with 400 to 500 face-to-face consultations a week. It is here that problems with players' accounts, and more often the complexities of the company's attempts to outlaw online trading, are dealt with when the in-game petitions and extensive phone support fail to provide a solution. Players face a ban if they're involved in online trading, and many come to appeal the ban in person, since the situation is sometimes simply a misunderstanding. I watch a young gamer explain his situation to the man at the counter. The consultant looks sympathetic, while the gamer looks weary and sad. The Kevlar-clad guard standing nearby doesn't seem to care and cracks open a can of Pocari Sweat, the delicious, if absurdly named, popular energy drink.

Whatever the nature of the misdemeanor the policy must, NCSoft insists, apply to everyone. As if by way of example, my guide disappears for a few minutes, returning to inform me that she had just found out that her own account had been suspended. She had, it seemed, been unfortunate enough to find some loot that had been black-marked by the gamemasters (GMs). She had chanced on a dropped rare item, known to the GMs to have recently been sold over a virtual trading market. Unaware of the danger and delighted with her find, our guide had picked it up and equipped her level 42 prophet. She hadn't been the one that bought it (that would have been utterly dishonorable), but when the virtual cops caught up with the item, it was found in her inventory. A quick check of server logs from the customer service boss, and it was concluded that she was not to blame. The problem account was reactivated. "Even though I work for NCSoft, I have to go through the same process as everyone else," laments the unfortunate young woman.

Onward to the telephone support rooms, where two women in their mid-30s, one heavily pregnant, tell me how proud they are of their team. The room is filled with young women, each one sitting at a computer and chattering into a headset. It could have been almost any call center in the world, except that everyone in this room has at least a level 40 character in either Lineage 1 or 2, sometimes both. The staff is mostly female because, I was informed, their feminine calm makes it easier to placate their mostly male, often rather irate customers. Their customers are, of course, the hundreds of thousands of Lineage players who log on to the servers every day across Korea. The games are so popular that there is, inevitably, some trouble. One of the worst aspects of this is the capacity for hackers and digital thieves to break into people's accounts, stealing their characters and looting their inventories. This kind of cybercrime isn't that common, but among an online population of hundreds of thousands, it becomes something that must be dealt with every day. More common are antisocial behavior and player killing, both of which are ubiquitous in the game but frowned upon when the players abuse their high-level powers. I am shown a prison within the game, a place where the GMs can incarcerate the unruly avatars in the most extreme of cases. "They can't escape from here," says an NCSoft GM, a plump girl in a frilly pink skirt. Ominously she turns round to confide: "None of these people can see me." I can only hope that she's referring to the game.

ZERG IDOLS

While RPGs attract many players from all walks of life, the real fanaticism is to be found in the competitive games that have made it onto TV. Early on a Friday night in Seoul's vast COEX mall complex, Interpreter Yang and I stand amid the television show audience at a regular Starcraft league game. One of five major leagues, and one of the key events for gaming television, the Ever Starleague has gathered an audience of about 500 people, with a few more standing outside watching the game on large screens. This is a small but regular occurrence, a far cry from the stadium events in which [RED]Nada Terran must defend his crown. For those events, says TV show manager In Ho Yoon, people will camp outside the night before, just so that they get the best place to sit when the thousands of fans file into the stadium. Vaguely mesmerized by the fast-paced RTS action, I am jolted from my reverie by a sudden roar from the assembled fans. Men and women, who a moment previously had been silently and studiously watching the buildup of Zerg and Marines, suddenly erupt into screams and chants. The chubbier of the two stone-faced players is in trouble and will soon concede defeat.

Ho Yoon says that Starcraft is now a self-perpetuating

phenomenon. In fact, it is to its benefit that the game is quite old, since the players who started out as the teenagers who first made it popular have subsequently grown up with the game and have inducted younger fans into its ways. With constant TV coverage, the game has continued presence in the Korean gamer's consciousness, and the lure of sponsorship for professional play means hordes of Koreans continue to want to get involved.

The games continue, and a striking young man in a white jumpsuit, apparently one of the country's best, attains an easy win. The crowd hoots and cheers. What is most important, says Ho Yoon, is the personality of the players he films. It is that personal appeal, along with their skill, that fills out the fan clubs and keeps the TV ratings high. I watch the blank-faced players preparing for the next match of the evening and am somehow reminded of a droning Nigel Mansell accepting the award for sports personality of the year. Still, it's what you do with your talent that really counts, and these guys (some of them dressed suspiciously like Formula One drivers) are superstars in their own right.

After the match my guides decide to go outside to enjoy a refreshing cigarette, and Ho Yoon asks me about gaming TV in the United Kingdom. Deleting Gamesmaster from my memory, I tell him about Time Commanders. He's not heard of the Total War games and says that they sound interesting. I agree but nevertheless feel slightly embarrassed—our fusty old war reenactment show doesn't exactly have a crowd of teenage girls perched attentively on the front row of the audience.

THE MIRROR WORLD

Back at NCSoft HQ, I am introduced to Y. H. Park, the lead designer of Lineage II. A stooping, smiley man in his late 30s, he bubbles with energy and explains effusively (in Korean) just how the game has fulfilled the expectations of both himself and his team. Lineage II is, in Asia at least, the benchmark game against which everything else is measured. A significant step onward and upward from Lineage (the game that initially made NCSoft popular and that has continued its own popularity), Lineage II provides greater scope and flexibility for a new generation of gamers. It has so far attracted 4 million subscribers across East Asia and regularly boasts 120,000 concurrent users in South Korea alone. On the world stage it is beaten only by Blizzard's phenomenal World of Warcraft, with the American company once again demonstrating that they are one of the few Western companies who are able to satisfy the tastes of the Korean palette.

Part of the reason for Lineage's success (aside from the ubiquitous marketing) is the need for Koreans to socialize and to distinguish themselves from their peers in what is (although heavily Westernized) still a fairly conformist society. It is perhaps because of this urge for socializing that player versus player combat also remains extremely important to the average Korean player. NCSoft is hoping that the arena combat of Guild Wars is going to provide the next generation of pro-gaming ambition, perhaps even supplanting the mighty Starcraft in its wider, TV-friendly appeal. It's been something of a success, but the full extent of its impact has yet to be made clear. When set against the backdrop of Starcraft, the budding RPG has a long way yet to go.

With so many players available to play online games, it seems as if Korea provides the ideal petri dish for such experiments in next-generation gaming. Yet there is a problem with this, and it finds its root in Y. H. Park's admission that no one really knows what the life span of MMOs really is, least of all in Korea. They will, he says, keep creating content for Lineage II for the foreseeable future but only in the form of new dungeons, new quests, and new items. The game will remain the same. And, when so many Koreans want to commit to a single game and not try anything very new or different for several years at a time, why should they change the formula? This kind of commitment means that the evolution of MMOs is likely to be slow, and perhaps less interesting, than that of other genres. When games like the now-ancient Ultima are still going, and still collecting cash, there seems little reason to push the games to new heights.

TROUBLE IN PARADISE

With a little more probing, it seems to some that not all aspects of the Korean game scene are unproblematic. For many gamers this is no Eden. For fans of Western games, the first-person shooters (FPS), strategies, and RPGs that are so popular in our part of the world, life is more difficult. Their interests lie outside the accepted Korean mainstream and, as such, get very little support. Despite critical acclaim, Half-Life 2 sold only moderately, and other games, such as the Knights of the Old Republic and other Bioware titles, have to be ordered from the United States, or pirated, if the gamers want to play them at all. One individual from the Korean games industry, who wishes to remain nameless for the sake of his continued employment, told me that he believes that the overbearing power of the major online game publishers in Korea ensures that any creativity or outside influence is stifled. The gigantic Net-gaming magazines, each one dealing exclusively with online games, which weigh in at over 400 pages each, almost fail to mention non-Korean games, and only World of Warcraft makes a significant dent in the editorial content of either magazine. The reason, of course, is money. Advertising keeps magazines and television channels afloat, and so they feel obliged to cover the games that pay for their advertising space. This, our confidant explained, makes the Korean scene dangerously narrow and victim to some very dull gaming. There is a third magazine, one that doesn't deal just with Korean Net-RPGs, but it's not nearly as influential as the hefty Net magazines or the free newspapers that report on events in Starcraft and Lineage. Worse, explained our contact, most of the online RPGs being pumped out of Korea are in pretty bad shape, with only a few companies like NCSoft being in any position to make them bug free and fun, as well as providing the support necessary to deal with player problems. People play many of these games, he laments, just to pass the time. "Little more than glorified chat programs" is a criticism of MMOs that we've heard many times over the last few years, and now, in this heartland of the MMO, it feels strange to hear those same words issue from Korean lips.

So the current climate can be depressing for Korean gamers who want to try something new and different. Even Counter-Strike: Source, initially popular in the PC baangs across the peninsula, has been virtually eradicated thanks to licensing troubles. The FPS genre has only the tiniest niche to play with and other game types even less so. Nor have consoles found much of a foothold, with the big publishers dissuaded by piracy or the forbidding nature of the Korean gaming scene. GTA3 has been made illegal by strict censorship laws, and even Starcraft has had to be neutered to avoid upsetting South Korea's sensitive authorities, red blood being replaced with black, and so on. It seems that most Western publishers don't have the cash, or the inclination, to break into the market and may never do so. Western visitors to Korea generally find the Kart Racer–playing, Starcraft-obsessed Koreans perplexing, and U.S. developers particularly have little or no idea how to breach the Korean gaming

ers within the Seoul area, that the capture or loss of a castle within Lineage II will regularly be celebrated or commiserated by a huge postgame meet-up. The clans routinely socialize in the baangs, and spectating some Starcraft matches after work is just another way to enjoy the company of like-minded people.

In Sook was the embodiment of what I'd come to expect from Korean gamers: someone deeply enamored with online gaming to the point where it defined her worldview and provided for her social life. She admitted that she was often at the computer a little too much (join the club) and that even gaming friends would complain at her to come out to PC baangs with them, but she felt that she had genuinely found a place to belong in her role as leader and organizer. Helping beginners in bimonthly sessions, or just hunting Lineage II's legions of monsters with her friends, had become a better hobby than she could ever have hoped for. Her opinions are echoed throughout Korean culture: Games are the best of pastimes, and if you can make friends while playing them, well so much the better.

But what about making money? In Sook said that she felt that virtual item trading was a bad thing and not at all in the spirit of true gaming. But it was, she acknowledged, a great shame that pro-gaming was limited to the likes of Starcraft. She would dearly love to play RPG games for a living. I nodded, echoing her sentiment. If Eve Online could be a real job, I suspect no one would ever see me or my gaming colleagues again.

As In Sook departed NCSoft's skyscraper office, I stood and looked out over the Seoul skyline—a teeming jungle of a city that seems on the brink of turning *Blade Runner* into a sunny reality. Truly, I mused, this is the crucible of a foreign gaming culture. The attitude of the place seemed at odds with the way that people approached gaming in the United

Kingdom, and yet there was nevertheless some shared vision, some inkling that we are, and have to be, akin in gaming. In *PC Gaming.* I found it hard to know whether perhaps, in some obscure way, the Koreans are simply tasting the future before the rest of us can catch up with them. Perhaps their adventures in the extremes of online gaming simply serve as an echo of a world that other cultures could one day find themselves immersed in, once broadband access is as common in our countries as it is in Korea. Perhaps it is a portrait of an alternate history, one that we could have experienced were it not for a few crucial differences. For good or bad, fun or tedium, this total commitment to multiplayer gaming is just one possible path along which Western gamers might one day tread. Or perhaps not. Perhaps we'll never see the like of it again.

But could this explosion of social gaming really be just an Asian phenomenon? Or could the whole world one day share in a transcontinental culture of massively multiplayer virtual worlds? Are we all stumbling toward some incredible Net-game future that lies beyond any present-day imagining?

Patiently digesting my seemingly deranged mumbling, Interpreter Yang simply shrugged. No one can really predict what will happen or what games we, or the Koreans, will be playing a few years from now. Who could have predicted anything of what had come to pass in Korea? "You know," said Yang, himself a veteran of Starcraft and Diablo II, "I don't really play much anymore, but my favorite game has always been Monkey Island."

"Yeah," I said. "I think some people back home like that one too."

Daniel Engber

Crying, While Eating

My sad, hungry climb to internet stardom

I found a picture of my girlfriend on a Japanese fetish site the other day. Yes, that was definitely her, cramming a piece of sausage into her mouth as tears streamed down her face. What's that right below her? A breast pump? This was all my fault. I'm the one who put that video online. They never told me that Internet celebrity would be like this.

A month before, I had signed up for a "contagious media" contest. The rules: Make a (nonpornographic) Web site. Promote it any way you want, short of paid advertisements. The page with the most visitors after three weeks wins.

The contest's host was Jonah Peretti, the creator of the much-forwarded Web site Black People Love Us. Peretti now runs a research group at New York's Eyebeam art and technology center that studies how sites get passed around the Internet. According to Eyebeam's experts, Web pages spread via the "Bored at Work Network"—the millions of shiftless desk jockeys whose fingers are glued to the forward button on their e-mail. Hoax product sites and pages that elicit a nervous laugh get passed around a lot, as do funny animal videos and movies of people dancing. But the most successful contagions are the oddballs, earnest amateurs like the *Peter Pan* guy and the *Star Wars* kid who had never tried to tap in to the Bored at Work Network. How could I compete with them?

As contestants, we had at least one advantage over the Peter Pan guy: a workshop that allowed us to kick around ideas with certified contagious-media professionals. Very few of us actually did. One guy announced his plan to create an animated dog that vomited things. After an awkward silence, the expert on hand suggested that he might want to think of a new idea. I was confused. Was a barfing dog any worse than the contagious Poke the Bunny site we'd learned about an hour before?

Forget the dogs and bunnies. I wanted my site to be about people, or food, or people and food. My friend and collaborator Casimir Nozkowski remembered a game he used to play at camp: Stuff some food in your mouth, and cry. We had our idea—Crying, While Eating.

On a rainy night, we drove around New York with a video camera, some sausage, a box of fried chicken, and an apple. I watched my friend Rob fast-forward through *Babe* until he got to the part where the sheepdog puppies are given away. Casimir zoomed in as Rob sobbed good, long sobs into the fried chicken. We took off a few minutes later with the sausage and the apple.

Crying, While Eating launched on a Thursday night with 12 videos. Christy, who was drinking a vanilla shake, cried because she was "good at lots of things but not great at anything." Tashi lamented the fact that "sex will never be that good again" while munching on Milano cookies. I ate buckwheat noodles with rooster sauce and blubbered about having "ruined Passover."

We waited until the next morning to send a batch of self-promotional e-mails. By the time we got out of bed, the blog Waxy had spotted our page on the contest Web site. From there, we got picked up by BoingBoing and Metafilter. I e-mailed the URL to a former co-worker in San Francisco that afternoon. He said he'd already gotten it from another friend in California, who had gotten the link from a guy in Austin, Texas. When I checked the stats that night, we had almost 50,000 visitors.

On Saturday morning, I got a message on my cell phone from "Joe," who claimed to be a marketing specialist in Los Angeles. "We have a deal in mind for you," he promised. When I called back, Joe said he'd seen Crying, While Eating on the "outrageous media" server and thought it was "fairly viral." He offered me a 60–40 split for placing ads on the site and asked if I was ready to "play ball." I made a counter-offer of 95–5, contingent on his telling me where he got my cell phone number. He didn't call back.

By the end of May, the site had gotten 7.5 million hits. Blog entries mentioning the site appeared in Dutch, Galician, Italian, Turkish, Norwegian, and German ("das ist doof"). People submitted videos from all over the world. Gwenda from Australia cried over "the shameful mistreatment of animals" while eating triple-chocolate ice cream. A guy from New Jersey sent footage of himself dressed up like a baby and crying over a plate of ribs.

Our egos swelled as we became D-list celebrities. An art gallery in New York requested videos for an upcoming exhibition, and a telecom company in Florida offered us thousands of dollars to put CwE clips in a commercial for long-distance service. Literary agents contacted us to discuss how the site could "make the jump to print." We got mentioned on VH1 and in *Entertainment Weekly* and were invited to appear on countless radio shows. Crying, While Eating even crossed over to the world of Internet porn. We got a huge number of referrals from a site called Goregasm ("where bones meet boners") and discovered that prospectors had snatched up the domain name www.cryingwhile masturbating.com. The sex-themed blog Fleshbot called CwE "our favorite new fetish of the year!"

And, yes, my girlfriend's video wound up on a Japanese sex site. Sure, that was a bit awkward, but I took some consolation in the fact that, after just two weeks, CwE was the top result of a Google search for "crying." I was a lock to win the $2,000 grand prize. I could make up for tossing my girlfriend to the Internet pervs by taking her out for a nice dinner. A really, really nice dinner.

Then my dot-com bubble burst. I'd been keeping an eye on a couple of our competitors, especially a video of people chugging Slurpees at 7-Eleven and a page featuring a masked man who freaked out to cell-phone ring tones. In a blink, a site I'd hardly noticed surged ahead in the standings. Forget-Me-Not Panties, a hoax page that offered futuristic, GPS-enabled chastity belts to concerned husbands and fathers, had become enormously popular overseas. (The Japanese in particular couldn't get enough.) Pretty soon, a Google search for "panties" led directly to their site. Crying, While Eating had dropped to third on the "crying" list, right below the Hungarian prog-rock band After Crying. We'd peaked too early, the contagious-media version of Howard Dean.

The contest ended a week later—with Crying trailing

Panties by more than 200,000 unique visitors. How had this happened? Hadn't anyone noticed the lovely write-ups in the *Ottawa Citizen* and the *Toronto Star*? Didn't anyone other than my parents watch us on *Best Week Ever*?

I pored over our traffic records to figure out what went wrong. Our television and radio spots hadn't really helped. All of that mainstream press came as we slid down from the contagious peak of our first few days. Newspaper articles didn't translate into lots of hits; all they did was lead to more print and television coverage. (The link I added to my *Slate* bio didn't help too much, either—it accounted for less than one-half of 1 percent of CwE's visitors.) Most of our traffic came from blog links and Web sites like College Humor and Something Awful.

It's easy to look back and see why Crying, While Eating did so well, at least for a time. It's a simple concept. It's interactive. It makes you laugh and feel uncomfortable at the same time. But there are two parts to contagious media. You have to make something that people want to spread around, but unless you're as lucky as the Star Wars kid you also have to do a little of the spreading yourself. CwE got lots of free publicity because it was an entry in a contest; if Casimir and I tried to make another contagious site, we'd have to do that legwork for ourselves. I don't know if we could pull it off. It seems like a real pain in the ass.

While the "Panty Raiders" took home the $2,000 jackpot, we did come away with two $1,000 awards. Crying, While Eating won Eyebeam's Alexa Prize as the first entrant to crack the Web's 20,000 most popular sites and the Creative Commons Prize as the most-visited site covered by a free distribution license. Best of all, I got to take home a humongous, four-foot-wide check. I thought about converting it into a coffee table, but I still owed my girlfriend a nice dinner. Now if I could only fit this thing through the front door at Nobu.

The contagious media sites cited in this article can be seen at the following URLs:

blackpeopleloveus.com
pixyland.org/peterpan/
ebaumsworld.com/starwarskidv.html
platinumgrit.com/pokethebunny.htm
cryingwhileeating.com
thebrainfreeze.com
ringtonedancer.contagiousmedia.com
forgetmenotpanties.com

Jay Dixit

Cats with 10 Lives

Why we need to regulate the cloning of felines and other animals.

Mary Ann Daniel and her husband, Roland, have had many cats, but Smokey, they said, "was one in a million." He showed up on Mary Ann's doorstep when he was just six months old. Roland's 11-year-old son had just died, and Smokey's arrival helped assuage their grief.

When Mary Ann read an article about a company called Genetic Savings & Clone that was offering cat cloning, she jumped at the chance. She took Smokey to the vet, who swabbed the cat's mouth in a simple, painless procedure to get a cell culture and shipped the cells off to Genetic Savings & Clone, where the company stored them in its gene bank.

Genetic Savings & Clone's feline cloning research project, Operation CopyCat, has already created a calico clone, and this summer, they used cells from a Bengal named Tahini to create two cloned kittens named Tabouli and Baba Ganoush. At $50,000 a clone, the service isn't cheap, but the company has five orders to be delivered this winter. The Daniels plan to order a clone of Smokey once the price has dropped.

Decisions like the Daniels' are controversial because,

many critics say, the company is taking advantage of the emotions of the bereaved. "It's a scam, really, because they're not going to get a replica of their animal, other than perhaps physically," said Peter Wood, a spokesman for People for the Ethical Treatment of Animals.

But a more interesting and poorly understood issue concerns the rights of Smokey 2.0. As emotionally complex an issue as cloning animals is, it's also dangerous, perhaps cruel and illegal, and almost totally unregulated. The consequences of remaking Smokey could be grim. Cloned animals often have severe health problems, and it's likely that Smokey 2.0 will die a grisly, early death. "Almost all clones will suffer and die, and they will do so not because of some natural illness or misfortune, but because researchers have chosen to bring them into existence using a process that is not understood well enough to use safely," wrote Hilary Bok, a philosophy professor at Johns Hopkins University. "These differences are not rare or anomalous: they are the norm."

Scientific studies have yielded a range of poor results. As reported in the journal *Nature Biotechnology,* 23 percent of all mammals that are cloned using traditional methods never reach healthy adulthood, instead falling victim to anemia, heart defects, liver fibrosis, obesity, and respiratory failure. The journal *Cloning and Stem Cells* reported that of 511 cloned pig embryos in an experiment, only 28 pigs came to term—of which a single clone was born healthy. One was born without an anus or a tail.

Nature Genetics reported that two-thirds of cloned mice died prematurely, and another survey found that they were much more likely to become obese in middle age. Eighty-five percent of embryos cloned from healthy cattle miscarried or had heart defects, joint problems, diabetes, severe anemia, or developmental problems. Even cloned cattle that appeared healthy scored lower on intelligence and attentiveness tests. The group that created Dolly the sheep, the first cloned mammal, had 276 failures before their success with Dolly. Dolly herself was euthanized in 2003 because of a lung tumor.

No one knows whether these problems are intrinsic to cloning or the result of human error in the use of cloning techniques. Most likely, the problems stem from the immaturity and complexity of the process. The most common cloning technique is nuclear transfer, in which scientists take adult cells from the animal they're trying to clone—in Smokey's case, the cells from his mouth—and insert them into eggs that have had their nuclei removed. This works because the DNA is the same for a cheek cell as for an embryo. But the procedure runs into problems because the cells also contain proteins that are required for their former adult functions. Once inserted into an egg, the cell has to get the message that it's supposed to become an entire animal, not just a mouth. Sometimes that message doesn't get through.

Genetic Savings & Clone says it is developing techniques superior to nuclear transfer. But the success of these new techniques remains to be seen. "It's true, to date, animals born to cloning have a higher incidence of health problems than animals born to natural reproduction," said Ben Carlson, a spokesman for Genetic Savings & Clone.

If cloned animals are likely to be born with severe health problems, is bringing an animal into existence to live a short, brutal life a form of animal torture? And if so, could a facility like Genetic Savings & Clone be found guilty under animal cruelty laws?

Animal cruelty laws prevent us from hurting animals without a good reason. The most important law is the Animal Welfare Act, passed in the 1960s after an article in *Life* prompted public concern about pet cats and dogs being mistreated, stolen, or sold to research facilities. The act specifies minimum standards of care for every sort of institution that houses animals, including zoos, circuses, puppy mills, and airlines.

State statutes approach the issue from the opposite angle, defining animal cruelty by specifying what is not permitted, namely, intentionally inflicting pain; killing an animal; or treating it with gross negligence, such as depriving the animal of food, water, and veterinary care. The severity of penalties varies by state. In 41 states, animal cruelty is a felony; in the other 9, it's a misdemeanor. Over the past two decades, laws have generally gotten tougher. When Senate majority leader Bill Frist was a student at Harvard Medical School in the early 1970s, he routinely, according to his autobiography, took cats home from animal shelters and practiced operating on them, often killing them. He wouldn't have faced harsh punishment, though, since that kind of animal cruelty was not a felony in Massachusetts then. But it is now.

Bringing a successful civil or criminal suit against a company or individual for causing pain or creating problems through cloning would require clearing several thresholds. First, you would have to show that it is the cloning company, rather than the veterinarians it partners with, that is working with the animals. "What we're doing here actually does not involve animals. It involves embryo manipulation and cell culture," said Ben Carlson.

The second hurdle concerns legal standing since the animals obviously can't bring the cases themselves. Most cruelty cases are brought by animal rights organizations, like the Humane Society, or by local prosecutors. But, so far, no organization has offered to bring cases on behalf of clones.

Third, for a felony conviction, you would have to show

that the cloner had intent to harm. State anticruelty statutes apply to intentional and malicious acts, like the one involving the Californian who, in a fit of road rage, scooped a bichon frise dog off a driver's lap and lobbed it into oncoming traffic (a crime for which the offender is now serving three years in prison). This standard exists partly because studies show that torturing animals can signal a dangerous pathology that leads to injuring human victims; the serial killer Jeffrey Dahmer impaled the heads of dogs and cats on sticks. But this standard also lets cloners off. Genetic Savings & Clone's specific intent is clearly not to hurt animals.

For misdemeanors, it's generally sufficient to show that a person knowingly condoned an act of cruelty. But again, demonstrating this would be hard. You might argue that the cloner knows that cloning techniques have a high failure rate and often cause serious problems. But so does reproduction in general. "In conventional breeding, anywhere around 30 percent of cats born don't survive," Carlson said. "Our losses are consistent with that."

If Smokey 2.0 drops dead from a strange cloning-related problem, Genetic Savings & Clone would likely face no legal consequences. "Cloning would not be covered by anticruelty statutes," said Taimie Bryant, a professor of law at UCLA. "As far as cruelty issues are concerned, cloning is below the legal radar."

Cloning animals for use as pets may or may not be ethical. And Genetic Savings & Clone may or may not be creating one Frankencat for every two successful Smokeys it sells. But if it were, the law is not equipped to deal with the problem.

Pet cloning exists in a regulatory no-man's-land. There are no laws about pet cloning, and there is no government agency that regulates it. The FDA has the authority to regulate cloned animals entering the food supply, but since cats are (presumably) not being cloned to be eaten, the agency has declined to get involved.

But a legal solution does exist, and it can be found by looking at the logic underpinning medical research standards. There, experimentation on animals is governed by the federal Animal Welfare Act, which requires that lab animals receive food and water, adequate veterinary care, enough room to turn around in their cages, and a humane death. Researchers can otherwise do just about anything to the animals—poison, irradiate, drown, or dissect them—so long as they do it for legitimate scientific purposes.

A similar standard could be applied to legislation regarding genetically modified animals, if society deems that a few botched Smokeys are worth the Daniels' happiness. Standards for care in the cloning facility should be prescribed, and cloning centers should be required to register with state governments.

In addition, a new category of law is needed for animals created through genetic manipulation—and for the losses that arise in the name of commerce, not science. The new category would include clones and other existing genetically modified animals, like goats that have rat genes so they'll produce low-fat milk; and enviropigs, whose manure doesn't stink.

In the meantime, the Daniels await the moment they can order their pet. "Knowing we'll get the clone is probably the only thing that keeps us going forward," says Mary Ann Daniel. "Smokey's unlike any other."

Well, except for his clone.

David A. Bell

The Bookless Future

What the Internet is doing to scholarship

Scenes from the Internet revolution in scholarship:

It is late at night, and I am at home, in my study, doing research for a book on the culture of war in Napoleonic Europe. In an old and dreary secondary source, I find an intriguing but fragmentary quotation from a newspaper that was briefly published in French-occupied Italy in the late 1790s. I want to read the entire article from which it came. As little as five years ago, doing this would have required a 40-mile trip from my home in Baltimore to the Library of Congress and some tedious wrestling with a microfiche machine. But now I step over to my computer, open up Internet Explorer, and click to the "digital library" of the French National Library. A few more clicks, and a facsimile copy of the newspaper issue in question is zooming out of my printer. Total time elapsed: two minutes.

It is the next day, and I am in a coffee shop on my university campus, writing a conference paper. A passage from Edmund Burke's *Letters on a Regicide Peace* comes to mind, but I can't remember the exact wording. Finding the passage, as little as five years ago, would have required going to the library, locating the book on the shelf (or not!), and paging through the text in search of the half-remembered material. Instead, on my laptop, I open Internet Explorer, connect to the wireless campus network, and type the words *Burke Letters Regicide Peace* into the Google search window. Seconds later, I have found the entire text online. I search for the words *armed doctrine* and up comes the quote. ("It is with an armed doctrine that we are at war. It has, by its essence, a faction of opinion, and of interest, and of enthusiasm, in every country.") Total time elapsed: less than one minute.

It is a few days later, and I am in my university office. I have seen a notice of a new book on Napoleonic propaganda and am eager to read it. A few years ago, I would have walked over to the library and checked the book out. But this particular book does not exist on paper. It is an "e-book," published on the Internet only. A few clicks, and the text duly appears on my computer screen. I start reading, but while the book is well written and informative, I find it remarkably hard to concentrate. I scroll back and forth, search for key words, and interrupt myself even more often than usual to refill my coffee cup, check my e-mail, check the news, rearrange files in my desk drawer. Eventually I get through the book and am glad to have done so. But a week later I find it remarkably hard to remember what I have read.

As these scenes suggest, in the past few years the world of scholarship in the humanities and social sciences has been astonishingly transformed by the new information technology. Above all, it has been transformed by the amount of source material now available online—some of it by paid subscription but much of it there for the taking by anyone with an Internet connection. Google made news in December with its ambitious plan to digitize the entire collections of several major research libraries (or at least the proportion that is in the public domain)—but to a much larger extent than the journalists who covered the story realized, the future that Google promises is already here. As I sit writing these words on my front porch, I can call up, in a matter of seconds, the sort of riches once found only in a handful of major research institutions: every issue ever printed of the *New York Times;* tens of thousands of classic and not-so-classic works of literature; a large majority of the books published in English before 1800; a million pages' worth of French Revolutionary pamphlets and newspapers; every issue of virtually every major American newspaper and magazine going back a decade or more; every page of most major American academic journals going back half a century; most major encyclopedias and dictionaries; all the major works of Western painters and sculptors. And much more is coming. Some of this material will remain available only in facsimile form. Much of it, though, is already entirely searchable. Name your keyword, and the Internet delivers the citations to you with the force of a fire hose in the face.

So far, most scholars have seen this transformation as a blessing—particularly those who do not have access to large, privileged research libraries. Indeed, its democratizing effects cannot be overestimated. Ten years ago, a historian whom I know took a job at the University of South Dakota. The entire library collection in her field ran little more than the length of her arm on the shelf, making real work on the subject effectively impossible, and she soon left. Today, a scholar in South Dakota, or Shanghai, or Albania—anywhere on Earth with an Internet connection—has a research library at his or her fingertips, even without access to the "subscription-only" content that makes up a large share of the holdings. The only protest I have seen against this democratization of information has come from Jean-

Noël Jeanneney, director of France's National Library. In a February Op-Ed piece in *Le Monde* that will long stand as a classic of unintentional Gallic self-parody, he complained that the Google project, by drawing principally on American libraries, would reinforce America's "crushing domination" of online information—no matter that the project will vastly expand the number of French books available as well and that nothing is stopping France from engaging in a similar project of its own.

But the Internet revolution is soon likely to become much more controversial, and for a simple reason: scholarship is fast moving toward a bookless future. Physical books are expensive to produce, and they are easily damaged or stolen. Shelf space costs money to build. Shelving and reshelving books costs more. Stacks have to be kept at the appropriate temperature and humidity; they need to be lit, cleaned, inspected, and insured. Why, it is already being asked, should universities pay large sums to preserve and circulate physical books if copies exist online? Just as physical card catalogs have been stored away or even destroyed, replaced by electronic ones, so physical books are likely to follow. Libraries, in turn, are likely to turn increasingly into virtual information retrieval centers, possibly located thousands of miles from the readers they serve. They already largely serve this function in the physical sciences, where the revolution in question took place much earlier and without much protest.

Writers such as Nicholson Baker, who eloquently objected to the disappearance of the physical card catalogs, are likely to greet this much larger change with despairing howls of anger. They will defend the physical book as an irreplaceable treasure, dwelling in covetous detail on every aspect of it: the paper, the typefaces, the binding. They will talk about its tactile pleasures, about the inimitable scent of dusty vellum and leather, and compare these things to the unnatural, unpleasant, uncomfortable experience of reading on a screen. They will cite the famous line of Borges: "I have always imagined that Paradise will be a kind of library." They will call the transformation another victory of soulless barbarism over true culture.

But this stance, for all its obvious aesthetic attractions, is far too sentimental and too easy. Not only is the advent of bookless or largely bookless libraries too large and powerful a change to be held back, it also offers too many real advantages for it to be considered a tragedy. Its democratizing potential, to begin with, counts for a great deal. Making vast libraries of learning available at no cost to anyone with an Internet connection is surely more important than preserving the rarefied pleasures of physical research libraries for those lucky or privileged enough to have easy access to them. The Internet also promises to make new forms of scholarship possible: new forms of research predicated on the rapid and efficient searching of vast databases, new hypertextual methods of presenting the results, and new means of ensuring their accuracy. Moreover, there are also ways—technological ways—of minimizing the aesthetic price to be paid.

What really matters, particularly at this early stage, is not to damn or to praise the eclipse of the paper book or the digital complication of its future but to ensure that it happens in the right way and to minimize the risks. For the risks are certainly real, and they go well beyond the disappearance of a particular physical object. The Internet revolution is changing not only what scholars read but also how they read—and if my own experience is any guide, it can easily make them into worse readers. Technological innovation can help to address this problem, and it is already beginning to do so. But it is not yet receiving the support it needs from either the publishing or the electronics industry.

II

How is the Internet changing the experience of reading? Consider the e-book that I found so hard to get through. Its title is *The Genesis of Napoleonic Propaganda, 1796–1799,* and in most ways it is a typical well-researched academic monograph. Fifteen years ago, its author, Wayne Hanley, would have easily found a university press willing to publish it as a sturdy hardcover volume with a print run of 500 or 1,000 copies.

But today specialized books of this sort are a distinctly endangered species. Their main purchasers—university libraries—have far less money to spend on these items than they once did. Computerized catalogs, subscription content, hugely expensive scientific journals, exploding storage costs: all these demands are putting tremendous pressure on budgets that are often already flat or declining. In response, libraries have cut back purchases or have started to form consortia with their neighbors, so that now only one research library in a given region may buy a particular book. As a result, specialized academic titles often sell as few as 200 copies, and university presses lose an average of more than $10,000 on each. The presses have cut back in turn, particularly in the more arcane precincts of scholarship. They are also passing the cost pressures on to those authors they do accept: it is becoming routine in some fields for university presses to demand subsidies of $5,000 or more to publish a book and to insist on strict limits on length. In some fields, the printed academic monograph seems dangerously close to extinction.

As scholars started to grapple with these problems several years ago, they concentrated, not surprisingly, on the immediate professional consequences: what happens to "publish or perish" if publishing becomes impossible? The obvious solution was to move specialized scholarship onto the Internet, but this presented its own set of professional problems. Today, anyone with a Web site is a "publisher." I have published several particularly specialized pieces of scholarship on my own Web site, for the sake of convenience. But this sort of "publishing" eliminates the peer reviewing that gives printed monographs the stamp of approval from the academic establishment, not to mention professional editing. Few scholars without tenure have the luxury to do it.

Just when these problems started to seem acute, Robert Darnton, a professor of history at Princeton, appeared on the scene with a suggestion. Darnton is a founding father of the field known as "the history of the book." (He was also my dissertation adviser.) Serving as president of the American Historical Association in 1999, he saw the chance not simply to write history but to make it. He proposed creating a new book award, called the Gutenberg-e Prize, in fields of history where the publishing crisis had grown particularly acute. The winners, instead of the usual certificate and check, would instead get their manuscripts "published" online. Columbia University Press came on board as a sponsor and to provide editing support. The result has been a "book series" well produced and prestigious enough to convince the most demanding tenure committee.

And Darnton had even greater ambitions. As he pointed out in a series of articles, electronic monographs can be much more than simple "books on a screen." He envisioned scholarship as hypertext, with "books" that would operate on several layers: a top layer of argument, from which readers could click down to a lower level of more detailed substantiation, and, below that, to further levels of raw evidence. Darnton himself provided an example in an impressive experimental article titled "An Early Information Society: News and the Media in Eighteenth-Century Paris," about the circulation of "seditious" information under the Old Regime. Published online (at www.indiana.edu/~ahr/darnton), it contains a 35-page text, illustrations, maps, a score of transcribed police reports, and 12 music files of seditious songs. An early modern society in the midst of one communications revolution (most notably, the rise of the newspaper) had come under study by a scholar using the experimental methods of another.

Internet publication can also improve scholarship in another way: by allowing for easy correction of mistakes. Last year, with much fanfare, an impressive new version of the British *Dictionary of National Biography* appeared, only to have various critics assail it for all manner of minor and not-so-minor errors. Making corrections easily available to users of the print version is a Sisyphean task, but correcting the online version is ridiculously easy. And where serious disagreements arise, the publishers can, if they choose, publish the debates themselves online. The result would be to make the work less an imposing, "definitive" monument and more an ongoing scholarly conversation—and that is an attractive proposition.

The Gutenberg-e series invented by Darnton now has 11 titles, ranging from Hanley's study of Napoleonic propaganda to Daniel Kowalsky's *Stalin and the Spanish Civil War* to Michael Katten's *Colonial Lists/Colonial Power: Identity Formation in Nineteenth-Century Telugu-Speaking India.* All are intelligent and lucid monographs, of interest principally to specialists. All take advantage of technology, even if they do not always live up to the promise of Darnton's hypertext model. Kowalsky's book includes short clips of Soviet newsreels alongside photographic illustrations but in such low screen resolution as to make them virtually unwatchable.

Gregory Brown's impressive monograph *A Field of Honor,* about French literary culture in the 18th century, has links to the collected works of several French authors, lengthy reproductions of archival documents, and hypertext links that allow one to move back and forth through the text in pursuit of particular themes.

For scholarly readers, these "books" are the shape of the future. Anyone who wants to check a citation in Edmund Burke can still find his works in print in any good library, but anyone interested in the Spanish Civil War who wants to learn Kowalsky's revisionist opinion of Soviet involvement, and anyone interested in the birth of modern literary culture who wants to consult Brown on the subject, has no choice but to read them online. And it is inevitable that great numbers of older, out-of-copyright titles will soon join these new ones in a cyberspace-only existence. One can almost hear the calculators clicking in the library offices: why keep multiple copies of *Hard Times, The Social Contract, Paradise Lost,* or *War and Peace* on expensive shelf space when anyone can download a perfectly good copy in his or her bedroom? Libraries that balked, decades ago, at putting much of their collections on microfilm, given the cumbersome machinery needed to read it, are showing no such hesitation when it comes to putting books online.

But again: what will this rush to cyberspace mean for the simple act of reading? This is where the problems with the Internet revolution are most obvious, and most harmful, as I discovered with Hanley's book and even more with Brown's *A Field of Honor.* Printed in standard form, even without the bibliography, this book would run 350 to 400 pages. It is clearly written, but it deals with difficult concepts, and it invokes dense and demanding theorists such as Bourdieu and Habermas. Even skipping certain sections, it took me many hours to get through, and by the end, the experience of reading on the screen had become, through no fault of the author's, distinctly painful.

III

Why is reading on the screen so genuinely unpleasant? Start with a basic point: Reading itself is a fundamentally unnatural act. Anyone who has ever taught a child to read will remember the difficulty involved in distinguishing, for instance, between lowercase *b, d,* and *p.* After all, if you move them around or flip them over, they are the same. We have to be taught to see them only as they appear, flat and unnatural, upon the printed page. And we have to be taught also to take in not just a few of these odd marks but the thousands that go into telling even the simplest children's story, to say nothing of the roughly 1 million that make up *A Field of Honor.* It takes years of practice before most of us do it easily, and even then, when it comes to difficult texts, it is the rare reader who perseveres, hour after hour, without a break.

Faced with these problems, Western culture long ago invented the optimal device for reading. Devised in the fourth century to replace cumbersome scrolls and parchments, it was called the book: a series of pages bound together between sturdy covers, light, portable, and easy to hold in the hand. Although some books, considered deserving of particular reverence, came to be produced in large "folio" formats that could be consulted comfortably only on a desk or book stand, most could be read virtually anywhere, in any position.

Remarkably, the great "printing revolution" that began with Johannes Gutenberg changed these practices very little. Gutenberg and his colleagues purposefully designed their new printed books as virtually exact physical copies of the manuscript books of the late Middle Ages. Put a printed book of the late 15th century side by side with a manuscript book of the same period, and it is surprisingly difficult to tell the Gothic typeface of the one from the scribal handwriting of the other. The "revolution" was a revolution in the means of production far more than in the nature of the product itself.

In this sense, our own communications revolution has been strikingly different from the earlier one. It has emphatically not been "Gutenberg II." To state the obvious: computer screens were not originally intended to replace books, and it is something of a technological accident that they are now coming to do so. Until the advent of the personal computer 25 years ago, computer screens were mostly used by professional computer programmers. They were modeled on earlier devices such as teletype terminals, with their typewriter keys and endless scrolls of yellow paper, which in turn had partially replaced punch cards and paper and magnetic tape. While screens were used to read programs, and data, and the early e-mail messages carried by Arpanet, very few people used them to read prose texts of any length.

This situation began to change with the rise of the PC in the 1980s, and the Internet a decade later, but still computer screens did not evolve very far toward the physical form of the book. Screen resolution improved, and today even a basic laptop screen will hold several hundred words in a reasonable facsimile of a printed page. Yet most screens remain wider than they are long, unlike printed book pages. Most computers make it easier to scroll down, line by line, than to page through a text. And screens are by no means as portable or as comfortable to hold as books. Personal digital assistants (PDAs), while more comfortable, display very little text at a time.

There are good reasons why an evolution toward the form of the book has not taken place. The wide screen that looks so unnatural for book reading is perfect for spreadsheets and for video. Scrolling down, line by line, remains the logical way to view things like computer code. And computers are designed above all for the comfortable input of information, which is to say that the screen is locked to a keyboard (or, in the case of a PDA or a Tablet PC, the screen itself becomes a slate designed for writing on with a stylus). In short, reading has remained distinctly subordinate to the computer's other uses. Nothing could be more different from the printing revolution, which had the reproduction of an existing form—the book—as its principal purpose.

Unfortunately, this subordination has grim consequences for reading. Start with the fact that what is already an unnatural task becomes more physically uncomfortable. One must stare at a screen in an upright chair or hold a heavy, awkward, and rigid piece of equipment on one's lap for hours. People will accept these constraints where no alternative is available—when working on a spreadsheet or playing computer games—but this is not the case with books. The relatively low resolution of even today's screens, compared with that of the printed page, tends to induce eyestrain. So does the fact that the eye remains at a constant distance from the screen. The tendency to scroll down rather than flip pages only makes things worse. It may seem a small detail, but a page becomes all the harder to concentrate on when the physical position of the words is constantly changing.

The very nature of the computer presents a different problem. If physical discomfort discourages the reading of texts sequentially, from start to finish, computers make it spectacularly easy to move through texts in other ways—in particular, by searching for particular pieces of information.

Reading in this strategic, targeted manner can feel empowering. Instead of surrendering to the organizing logic of the book you are reading, you can approach it with your own questions and glean precisely what you want from it. You are the master, not some dead author. And this is precisely where the greatest dangers lie, because when reading, you should not be the master. Information is not knowledge; searching is not reading; and surrendering to the organizing logic of a book is, after all, the way one learns.

If my own experience is any guide, "search-driven" reading can make for depressingly sloppy scholarship. Recently, I decided to examine the way in which the radical 18th-century thinker d'Holbach discussed warfare. I could have read his book *Universal Morality* in the rare-book room of my university library, but I decided instead to download a copy (it took about two minutes). And then, faced with a text hundreds of pages long, instead of reading from start to finish, I searched for the words "war" and "peace." I found a great many juicy quotations, which I conveniently cut and pasted directly into my notes. But at the end, I had very little idea of why d'Holbach had written his book in the first place. If I had had to read the physical book, I could still have skimmed, cut, and pasted, but I would have been forced to confront the text as a whole at some basic level. The computer encouraged me to read in exactly the wrong way, leaving me with little but a series of disembodied passages.

Of course, there was an obvious alternative to reading on the screen: printing the thing out. With d'Holbach, I did print the first 100 or so pages, only to have my computer chirpily announce that it was time for another expensive ink cartridge. Printing out is an expensive proposition, as well as a troublesome and time-consuming one. In any case, printing a book out goes against the point of using the Internet in the first place. Printing takes away the hypertext, multimedia functions built into works such as Darnton's. And reading on the screen, frustrating as it is, has certain advantages. On my own computer I keep several foreign-language dictionaries and a good thesaurus; the *Oxford English Dictionary* and the *Encyclopædia Britannica* are just a few clicks away. I can have several books open at the same time to compare texts. I have immediate access not just to Internet resources but, on my hard drive, to just about every note I have taken and every piece of writing I have done in the last 20 years. Finally, there is a certain intellectual justification for instant gratification: ideas occur with particular readiness when you can pursue a train of thought quickly from one book to another. Readers of physical books have long known this form of research—it is called browsing in the stacks.

IV

Is there a way to have these advantages without doing lasting harm to the experience of reading itself? Perhaps, at least in part. What is needed is a technological solution, in the spirit of the original Gutenberg revolution, the revolution of the 15th century. That is to say, what is needed is a computer that looks and feels exactly like a book. And it is coming. Recent advances in "electronic ink" and new reading devices so far sold only in Japan come tantalizingly close to this ideal, but there are still major obstacles on the road—not all of them technological.

To date, the various attempts to produce specialized electronic reading devices have mostly been failures. In 1999, at the height of the tech boom, gadgets called the SoftBook and the Rocket eBook created a brief stir when they came on the market. Designed explicitly for reading, they were light, easy to hold, and had vertical, pagelike screens, although with poor resolution by today's standards. But both devices flopped. So did a cheaper, smaller, PDA-like version called the eBookman, sold by Franklin.

The electronics industry has had marginally greater success getting people to use their PDAs, laptops, cell phones, and Tablet PCs for reading. A number of programs such as Palm Reader, Microsoft Reader, and Acrobat eBook make these computers as "booklike" as possible, including special screen fonts that in theory reduce eyestrain. But e-books have not yet come close to challenging the hegemony of printed books. Their sales, while growing, amount to a tiny percentage of industry totals (just under $10 million in the first three quarters of 2004). Barnes and Noble, which made a significant effort to sell e-books on its Web site, quietly discontinued the practice last year.

Perhaps the surest sign of the insignificance of e-books is that for years electronic versions of best-sellers have been available on file-sharing services such as Kazaa without causing much scandal or even notice. The *New York Times* estimated recently that as many as 25,000 titles can be downloaded, including all the Harry Potter novels and *The Da Vinci Code*—but sales of the print versions have not been hurt enough to make the publishing industry worry. Most book editors I know are not even aware of the files' existence.

The physical e-book readers have failed for three reasons. First, and most important, the various devices are simply not booklike enough. While the specialized reading machines, PDAs, and cell phones are lighter and more comfortable to hold than computers, their screens are terribly small and coarse. A Pocket PC screen using Microsoft Reader can display barely 100 words at a time, which makes even a relatively short book more than 1,000 screens long. The Tablet PC does better on this score, but it is heavy, rigid, and awkward and thus has been marketed almost exclusively as a note-taking device. Nor does any of these gadgets really solve the eyestrain problem—even with Reader, which Microsoft released to considerable fanfare a few years ago and has since allowed to wither. Secondly, the specialized reading devices, while slightly more comfortable, were too expensive. Who was going to spend $800 on RCA's color version of the Rocket eBook when a few hundred dollars more would purchase a full-featured laptop with a better screen?

Most importantly, the companies, clearly fearing that the devices themselves would not generate sufficient income, focused instead on selling "proprietary content"—that is, encoded versions of books under copyright. Several of them, such as the SoftBook, initially did not even provide a way for readers to load their own readings onto the devices. As might have been predicted, the companies thereby drove themselves into a classic vicious circle: publishers refused to make more than a handful of titles available without evidence of readers' interest, and readers, faced with a tiny selection of titles, shunned the devices entirely.

This story points to one of the most powerful factors inhibiting the development of booklike computers and reading devices: the publishing industry itself. For the moment the sharing of pirated book files over the Internet has attracted little attention. But imagine the development of a computer that really was as easy and as comfortable to read as a book. Would book sharing become as great a threat to publishing as music sharing has been to the record labels? True, there is no real textual equivalent to the "ripping" of a music CD. Most readers have little incentive to turn libraries of books they have already read into shareable files, and doing so is far more difficult than ripping a CD. It involves either breaking open a coded file or tediously scanning a book, page by page.

Still, only one person has to take the trouble, and within hours millions of copies can be circulating on the Internet. Remember that text files are very, very small compared with music or video. A best-seller can be downloaded over a high-speed connection in a matter of seconds. This scenario must cause publishers some sleepless nights.

And the moment may be coming closer. In Japan, Sony and Panasonic recently released new-generation reading devices that put clunky predecessors like the SoftBook to shame. Sony's entry, called the LIBRIé, is particularly impressive, for it employs a technology called "electronic ink," in which the screen is composed of tiny "microcapsules" that can turn black or white through the manipulation of an electronic field. While previous screen technologies required an internal source of illumination, electronic ink does not, making it easily readable even in full sunlight and cutting back significantly on bulk and power consumption. It also has greater resolution than previously achieved. The LIBRIé weighs only a little more than a pound, can run for weeks on ordinary AAA batteries, and displays half a million pixels on its six-inch screen—six times more than most PDAs. To put things simply, it weighs the same as a book and looks very much like paper (although it takes a frustratingly long time to "turn" pages). For the moment, it is available only in black and white, but full-color versions are said to be only a few years down the road. And even the first-generation model costs less than $400.

But will the LIBRIé succeed? At first, Sony and Panasonic both repeated the disastrous strategy of allowing only proprietary content, downloaded from special Web sites for a fee, onto the devices, and Japanese publishers refused to make more than a relative handful of titles available. Will people pay hundreds of dollars for a reading device when they cannot use it to read work documents or free books downloaded from Web sites of their own choice? Will publishers make enough books available to persuade readers to purchase such a limited device? I have my doubts. More recently, both companies have made it possible to use the devices to read other documents, but only after a complicated conversion process that will repel most users. Sony has yet to announce a release date for the LIBRIé in the United States. I suspect that such devices will only truly succeed when they have the full capacities of computers—so that readers can download Web pages and electronic books onto them as easily as I now download my research materials onto my laptop.

V

When this happens, it is entirely possible that a second Gutenberg revolution will finally take place, bringing about the long-discussed paperless office, together with the bookless library. If I had an inexpensive, full-function computer that was roughly the size and the weight of a hardcover novel, with a high-resolution, paperlike color screen, a detachable keyboard, and wireless Internet access, I would be quite happy to stop squeezing new bookshelves into my basement and office.

But this scenario is not inevitable. The publishing industry can do a great deal to frustrate it by refusing to make copyrighted material easily available in electronic form for fear of piracy. Traditional book lovers can also do a certain amount to frustrate it by stigmatizing electronic publishing as the sign of a second dark ages. And if demand for advanced reading devices remains low as a result, then the electronics industry will not invest significant resources in electronic ink, and the LIBRIé will go the way of the SoftBook.

The traditionalists may applaud this outcome, but they would be wrong to do so. Frustrating the development of real "booklike" reading devices will undoubtedly slow the transformation of libraries into virtual information centers. It will slow the pace at which books are scanned and then relegated to subbasement storage facilities. It will stave off the death of the academic monograph. But it will not stop any of these things, not least because the financial pressures bringing them about are too strong. It will just make the electronic books we have—and we will have more and more of them—unnecessarily awkward and difficult to read. It will encourage searching rather than true reading and turn eyestrain into a new form of occupational hazard for scholars everywhere.

It would be far better for publishers to learn from the semidisastrous experience of the music industry. Threatened by Napster and its clones, the record labels initially tried to shut down the new technology by heavy-handed legal tactics but eventually made songs available online themselves for a reasonable price and with reasonable restrictions. And when they did, consumers flocked to services such as Apple's iTunes. The publishing industry would do well actively to plan for the day when it will sell a majority of its products not on paper but over the Internet, to consumers who will read them on new, attractive, paperlike screens.

But what will happen to the experience of reading, particularly in scholarship? Will traditional reading—the slow, serious reading of entire texts—sink from sight in an ocean of hypertext searching? We can at least hope not. Reading itself will surely change, for the simple reason that new electronic devices, even if they look and feel exactly like books, will still be different from them—far more different than Gutenberg's books were from their printed predecessors. For one thing, they will most likely be full-function computers, with word-processing and Internet capability. But we can hope that as the physical discomforts and frustrations of reading on a screen diminish, more traditional sorts of reading will find their way into cyberspace—that readers, holding a truly "readable" computer in their hands, will not abandon themselves to searching and clicking but will instead find it comfortable to sit, and read slowly, and stop to ponder what they have read.

And even the newer forms of reading are not to be entirely deplored. For a start, they will encourage new works of scholarship to take full advantage of the possibilities of hypertext and multimedia in a way that the pioneering Gutenberg-e books, for the most part, have not. Even more important, they will raise the simultaneously glorious and terrifying possibility of having an entire world library at one's fingertips. In any case, we need not assume that one form of reading will entirely replace others. Different forms have always coexisted with one another. Before Gutenberg, when books were rare and expensive, the dominant form was probably the slow, intensive, repetitive study of sacred and quasi-sacred texts. It may have remained so even during the first centuries of printing, but gradually it was challenged by more "extensive" sorts of reading, involving the relatively quick, onetime perusal of books for entertainment and the speedy acquisition of information. But the first sort never disappeared, and of course it still exists in many settings. Now, with the Internet, have come yet newer styles. But they, too, can coexist with older varieties, especially within the academy.

Perhaps this is too sanguine a view. But scholars are,

after all, professional readers. The books that they read are likely, as time goes on, to have a physical existence only as evanescent electrical patterns on delicate pieces of machinery, and this technology will affect the way they read. As long as the things they read are as physically easy to read as paper books, scholars need not be overwhelmed by their new world of choices, any more than the scholars of the Renaissance were overwhelmed when faced with the sudden explosion of books brought by the printing press itself (although they certainly had to invent new strategies to deal with it). Those scholars adapted and flourished, and so can we. The bookless future need not be a barbarian age. The character of our culture will finally be determined in the old way, not by the form of words and ideas but by their content.

David McNeill

Mr. Song and Dance Man

Daisuke Inoue taught the world to sing with the karaoke machine but never bothered to patent it, losing his chance to become one of Japan's richest men. Is he bitter?

For a man who lost out on one of music's biggest paychecks, Daisuke Inoue is in fine form: toothy smile spreading over the big, rough-hewn face of a natural comedian.

The good humor comes in useful for interviews like this when he is inevitably asked whether he regrets not patenting the world's first karaoke machine, which he invented in 1971.

After 34 years, during which his unlikely contraption has conquered every corner of the globe, accompanied by the sound of a billion strangled, drink-sodden, earthling voices, the question must sound like the whistling of an approaching bomb. But the smile stays.

"I'm not an inventor," says the 65-year-old in his small Osaka office. "I simply put things that already exist together, which is completely different. I took a car stereo, a coin box, and a small amp to make the karaoke. Who would even consider patenting something like that?"

"Some people say he lost 150 million dollars," says

Inoue's friend and local academic Robert Scott Field. "If it was me I'd be crying in the corner, but he's a happy guy who loves people. I think it blows his mind to find that he has touched so many people's lives."

In Japan, many now know, thanks to TV specials and a new movie biopic, that Inoue was a rhythmically challenged drummer in a dodgy Kobe covers band when he hit on the idea of prerecording his own backing tracks.

The band had spent years learning how to make drunken businessmen sound in tune by following rather than leading and drowning out the worst of the damage, so Inoue knew the tricks of the trade when the boss of a steel firm asked him to record a tape for a company trip to a hot springs resort.

Unbeknownst to millions of once-peaceful pubs, karaoke had been born. Inoue and his friends gave it a leg up into the world by making more tapes and leasing machines to bars around Kobe. Taking the machines for a spin cost 100 yen a pop—the price of three or four drinks in 1971—and Inoue never thought it would make it out of the city.

By the 1980s, karaoke was one of the few words that required no translation across much of Asia. Communist China embraced it, and Hong Kong sent it back to Japan as karaoke box, small booths where friends and family could torture each other in soundproofed bliss.

Inoue languished for years in international obscurity. But in 1999, after karaoke had stomped noisily into the United States and Europe, *Time* magazine astonishingly called him one of the 20th century's most influential Asians, saying he "had helped to liberate legions of the once unvoiced: as much as Mao Zedong or Mohandas Gandhi changed Asian days, Inoue transformed its nights."

"Nobody was as surprised as me," he says.

Last year, he was awarded the Ig Noble Peace Prize at Harvard University, a jokey award presented by real Nobel winners, receiving a standing ovation after calling himself the "last samurai" and attempting a wobbly version of the 1970s Coca Cola anthem "I'd Like to Teach the World to Sing."

The Nobel laureates in turn (or in revenge) murdered the Andy Williams standard "Can't Take My Eyes off You." Inoue loved it, laughing throughout. "I wish I spoke English," he says. "It would make life easier, and I could go to the US again, do speaking tours and make some money."

Now he is the subject of a new fictionalized movie account of his life, called simply *Karaoke* and directed by Hiroyuki Tsuji, currently on release in Japan and starring an actor considerably better-looking than the weathered, plump drummer of 1971. "At least they got someone tall to play me," he laughs.

A typical Osaka businessman, amiable, fast talking, and with a slightly untamable air, Inoue once tried working in a proper company but balked at wearing the salaryman's uniform: the dark pinstripe suit. "I looked like a rocker, and it didn't go down very well."

He didn't use a karaoke machine until he was 59 but loves to listen to syrupy pre-1960s ballads; his favorite English songs are "Love Is a Many Splendored Thing" and Ray Charles's "I Can't Stop Loving You." "They're easy to sing, which is good because I can't sing at all."

Inoue is tormented by daft questions but takes them in his stride. "People approach me all the time and ask me if I can't help their husbands sing better, and I always say the same thing. If the singer was any good, he would be a pro and making a living at it. He's bad because he's like the rest of us. So we might as well just sit back and enjoy it."

These days he makes a living selling, among other

things, an eco-friendly detergent and a cockroach repellent for karaoke machines. "Cockroaches get inside the machines and build nests, chew on the wires," he says. Friends say he is the ideas man, while his wife, who works in the same Osaka office, helps bring them to life.

In the 1980s, he was running a company that successfully managed to persuade dozens of small production companies to lease songs for eight-track karaoke machines. But in the early 1990s, laser and dial-up technology left the firm behind; bored and depressed he had a breakdown but bounced back to life thanks to his dog. "I had to look after it, and it got me out of the house." He says his next business venture will involve dogs.

Not everyone thanks Inoue for his invention. A 2004 black comedy called *Karaoke Terror* depicts a bunch of bored, middle-aged women and a group of college kids, both obsessed with karaoke, who go to war with each other, destroying a whole city: karaoke as an almost too-easy metaphor for the emptiness of contemporary Japanese culture.

But the ponytailed businessman believes the little box he put together in Kobe has done far more good than harm. "As something that improves the mood, and helps people who hate each other lighten up, it has had a huge social impact, especially in Japan. Japanese people are shy, but at weddings and company get-togethers, the karaoke comes out and people drink a little and relax. It breaks the ice.

"It's used for treating depression and loneliness. Go to old people's homes and hospitals around the country and there is a karaoke machine. I keep hearing of places where karaoke is huge—like Russia—and it is used as therapy. It makes people happy everywhere. When I see the happy faces of people singing karaoke, I'm delighted."

The biopic supports the idea that karaoke is socially useful, rather than the bane of quiet pint drinkers everywhere. Kicking off with a grim list of suicide statistics among middle-aged Japanese men, it depicts a salaryman losing his job, wife, and son after he is fired. He starts singing karaoke and finds a new purpose in life.

"We went to see the movie together," remembers Robert Scott Field. "They got these good-looking actor and actress to play him and his wife, so he was really happy. Afterwards he said: 'I get letters and e-mails from all over the world, and now they've made a movie of my life story. I have to pinch myself. You can't buy things like that.'"

Koranteng Ofosu-Amaah

Cultural Sensitivity in Technology

Everything is local.

The "Coltrane" release of Lotus Freelance Graphics, the 1998 version of the presentation graphics program bundled in Lotus SmartSuite, was famously held up for a month when it was found out that one of the standard clip art images used in the out-of-the-box presentations had a tiny 20-pixel image of Taiwanese currency (rather than mainland Chinese currency). Or was it the other way round? An eagle-eyed quality assurance engineer discerned the offending image. Needless to say, this was something that could not only cause offense but also affect purchasing decisions in the involved markets. Productivity suites like SmartSuite are very profitable products since they are among the indispensable tools in every office; everyone needs word processors, spreadsheets, organizers, and presentations.

I was a little canary in that Freelance mineshaft (or was it a minefield?) when a couple of months earlier in the project I discovered a seriously outdated map of Africa while integrating new clip art into the product. I wrote one of the "software problem reports" or "sprs" of which I am most proud; it was entitled something like "Upper Volta should be Burkina Faso." The spr also made a brief mention of the fact that, as far as I could tell, there were similar problems in our maps of the former Soviet Republic since many of the states' names had changed earlier in the decade. I believe that my spr was deferred to a later release; it was too late in the product cycle, and we'd have had to go back to the company we licensed the art from, and so forth. As it turned out however, the China/Taiwan currency error caused the entire suite to be delayed while the test team painstakingly examined the clip art and other areas of the product, vetting for similar outrages that could endanger sales.

Incidentally, the acronym spr is a distinctively Lotus term. After a decade of indoctrination, I continue to use it rather than the more clinical "defect," which is the favored term at IBM, or "bug," which is the more widely used term generally. In fact, you can still tell whether a Lotus employee has the "old Lotus" DNA or the "new Lotus/IBM" variety by the terminology they reflexively use.

As a further aside, the Lotus SPR database is actually one of the most successful and useful applications of Lotus Notes technology ever, and, in a similar manner, the Bugzilla project is one of the best things to have sprung out of the Mozilla effort. Certainly it was more useful than the Mozilla Gecko browser suite (mother to the Firefox) until the M7 milestone, the Mozilla 0.7 release, was reached years ago. This suggests that a new kind of software law is at work here, one analogous to Zawinski's Law* on "software expanding until it can support email" (and now feed technology). According to this law, which I'll call Koranteng's first law of software systems:

* Jamie Zawinski coined his law of software envelopment as follows: "Every program attempts to expand until it can read mail. Those programs which cannot so expand are replaced by ones which can."

> A software platform reaches its tipping point once it can serve as a bug reporting system.

Intuitively this makes sense: Once developers can view, file, and retrieve bug reports using their own products, they will be more likely to use them, and their confidence in their mission will improve dramatically.

But back now to my original topic. I'm sure that the artists who created the clip art in that design firm had no subversive intent; they probably just used outdated stock art as their source material. Still, you sometimes wonder at these things, and, in this vein, I'm reminded of the Nobel Prize–winning Portuguese author Jose Saramago's wonderful novel, *The History of the Siege of Lisbon,* in which a copy editor proofreading a historical novel defiantly changes a crucial word in the text and in this way also changes the outcome of a famous historical battle from a miserable defeat to victory. A masterful alternate history of Portugal then develops, and its culture is fully reimagined. I'm sure that bored developers sometimes slip similar things into their products, and not just serendipitous "Easter Eggs." In the same release of Freelance, we put together a hidden game and a screen show with the photos of all the developers that could only be reached by activating an unusual key sequence (I won't kiss and tell here). These days, there are Web sites that catalog the hidden secrets (and occasional hidden features) that can be found in many software products.

More seriously though, the Freelance incident (or similar incidents like those involving the release of new dictionaries and thesauruses from Microsoft) goes beyond the cultural sensitivity with which this post is concerned into the rarefied realm of political sensitivity. For example, if you typed the phrase "a fool," the Microsoft Word 2000 thesaurus would offer "trick" as a replacement. The opposite was the case when typing "zzzz" in Word 97, in which case the thesaurus suggested "sex." Later service packs had to be issued to "fix" these and other glitches. This is also the case with product names: You don't want the mere mention of your product to cause snickering or give offense. A misconceived name or culturally insensitive feature can be a black eye that will embarrass and divert revenue from even the highest-flying company. In 1998, Lotus SmartSuite was a $400 million a year business; that month's lost revenue was no small thing.

FORENSIC INVESTIGATIONS

Although obviously important, the task of internationalization, which aims to make technology available in different languages and scripts, is a mostly thankless, and oft-neglected, one. From a developer's standpoint, you think about it mostly in terms of dealing with "resource bundles." Wherever you find onscreen text, you have to replace it with a key word and put the text in a file, along with the key word to reference it. This file is then sent off to translators for each of the different languages you support. The resulting bundle of translated files is packaged in the released program. When the software runs, it figures out the user's language, looks up the correct language file, and uses the given key word to obtain the resource string for display.

Designing software in this way allows you to support new languages by simply dropping a text file for a new language in the right location. This is a pain because you probably started with a little prototype of a user interface and now you're being asked to find all those hard-coded strings in the user interface and "do the right thing." It is also confusing because, from then on, you see key words instead of phrases in the dialogues in your development environment.

Adding to this irritation is the fact that internationalization typically only becomes an issue late in the game, when the really interesting work is over and you're ready to think about the next big thing. Instead, you're grudgingly forced to worry about handling different writing systems; dealing with bidirectional text; or (that old favorite) wrestling with the "special characters" in the wide variety of programming languages, file formats, protocols or operating systems in your software. In computer science there is a general category of problems that have to do with delimiter characters (underscores, hyphens, colons, semicolons, angle brackets, and other special characters). Software is very concerned about structure, and the parsers of our Tower of Babel file formats need to know where records begin and end. Every tribe in the software world has its idiosyncratic ideas about language, and which characters to use to separate records, as it seeks to impose structure on the world. There are ever-shifting alliances, and fashions, and it is a heady proposition to make sense of things as the tribes interact. Those who excel at these tasks are akin to forensic investigators: patient, precise, and possessed of a keen eye for detail.

The consequences of overlooking these seemingly small details are, again, larger than one might expect. If, for example, your software mangles names with accents, you'll have real trouble selling in France. The same applies to ampersands and apostrophes—which have significance in SGML (standard generalized markup language) and its derivatives (HTML and XML). And don't get me started on wider character set and encoding issues. As an ironic case in point, while developing Lotus K-station, IBM's first attempt at a portal, one of our best business partners was somewhat stymied in his development and extraordinary evangelism of our product because, in its earliest release, Lotus K-station couldn't handle the ampersand in his company's name.

Luckily for us, he temporarily renamed his organization in his corporate LDAP directory; Sun & Son became Sun and Son until we worked out the quirks. To this day, I always make sure to test whatever product I work on with its organization name. It's surprising though, how often this kind of problem recurs.

THE SCENE AT HOME

In the Internet era, most technology, and certainly all software development, has to have global concerns in mind. It is said that the sun doesn't set on an IBM project, and it is true that I work with a very diverse set of colleagues the world over. Presumably, one of the benefits of such a widely dispersed and diverse workforce would be to mitigate the likelihood of issues in this area. This benefit, however, is only realized if everyone gets the opportunity to give their input. As my current project has been going through translation and localization testing of late, I've been thinking a lot about the different strategies for handling internationalization.

The "old Lotus" process, for example, was fairly decentralized: Each product group would be assigned an internationalization team. In addition to being localization gurus, members of such teams were domain experts and knew the product inside out. Having the team involved from the very beginning in the product development cycle had many benefits since it enabled them to give crucial design feedback very early and iteratively. Your first prototype was immediately critiqued from their standpoint. The obvious downside of decentralization was the lack of uniform standards, chaos that our translation teams couldn't bear. Some products were very easy to localize; others were, to put it mildly, far less so. This was in keeping with the culture of Lotus, which was historically full of artists, writers, historians, sociologists, physicists, and folks like me: electrical engineers who picked up software engineering almost as a hobby.

As we transitioned to the more centralized IBM globalization process, we got the benefits of uniform standards and greater resources. More languages could be supported in the initial release, and you could at least point developers to documentation about the processes they should follow and in that way avoid the usual ad hoc stumbling about. The IBM style is all about corporate professionalism; it is more ponderous and process minded: hence the "Big Blue" nickname. There are many benefits to organized processes, but they can also come at the expense of domain expertise, sensitivity, and a much faster feedback loop. Members of the test teams are generalists and, on any given week, will be testing a wide variety of products. They often aren't aware of the nuances of your particular product and are often barely getting up to speed with it by the end of the testing phase. Hence localization issues frequently surface too late in the game and cause unnecessary reworking and product delays. These processes are of course constantly being tweaked, as all processes are, and in recent years, I've seen a renewed emphasis on domain expertise in the internationalization teams that has improved our product development considerably.

One of the lessons here may be that software engineering is very different from computer science, precisely because of its greater preoccupation with what Graham Greene called "the human factor." For once you bring people into the picture, you bring culture: conversations, marketplaces, attitudes, details, conventional wisdom, and sometimes blind spots. Working on the Web exposes one to a multitude of such voices, and, for that reason, I view it as a conversational puzzle. The clues are sometimes irreverent and cacophonous, but the solutions are always interesting and, once found, ultimately rewarding.

Here is a brief example, which should help to tease out the kind of technical, design, and business dilemmas that can arise if you aren't attentive to the cultural issues in your product development.

I embarked over the past couple of weekends on a mass digitization project and scanned, retouched, and uploaded 2,000 or so old photos from shoe boxes under my bed. The technology involved in this exercise was scanner hardware and image acquisition software, the bundled Adobe Photoshop Elements for color adjustment and red-eye correction, and a couple of online photo-sharing services: Yahoo Photos and Flickr (coincidentally, I started the day Yahoo's acquisition of Flickr was announced).

I noticed very quickly that all the photos that I uploaded to Yahoo Photos had somehow turned out darker than on Flickr. Both services resize uploaded photos; when you reduce the size of images, you have to select the pixels and colors you are going to use, but the photo-resizing algorithm used by Yahoo Photos was giving worse results. This was noticeable to me because a large number of photos featured darker-skinned people like me. The originals looked fine on the screen and wherever there were lighter skin tones. In the case of darker skin tones, however, the resized photos were not so good. This meant that if I didn't believe in the virtues of Save Lots of Copies Everywhere (SLOCE), I would have leaned toward Flickr and stopped using Yahoo Photos.

I also noticed that my experience of the Flickr Web site was very different from that of my family and friends who used Internet Explorer. Almost all of them complained immediately about the first bunch of photos I presented to them. My initial thought was that the problems stemmed from the different browsers we were using (Internet

Explorer in their case, Mozilla in mine). After a little investigation, however, I found that the real reason for the complaints was the Flash plug-in. If you had Flash installed, Flickr was coded to use it to display images. On the other hand, if you didn't have Flash, the browser's native rendering took over the display task. It turns out that images that are rendered in the Flash plug-in have a slightly darker tinge than the images rendered directly by the browser itself. This is not normally noticeable unless darker skin tones are involved, as was the case here. This problem became even worse when they tried the screen show feature. The black background of a Flickr screen show (also implemented in Flash) impaired the contrast still further.

Finally, it became clear while I was retouching the photos, and constantly forced to tweak them manually, that the Quick Fix and Auto Correct options in Photoshop were also better suited for lighter skin tones. Now, this tweaking is not a big deal in the case of a few photos—indeed it's fun to fiddle with photos. But after a couple of hundred images, it gets tiresome. I found myself longing for "smarter" recognition by the software or, at least, for a nice "dark skin" option that I could set in a preferences dialogue. In short, I started to think about abandoning Photoshop for a different program.

I mention these nitpicks with otherwise excellent and useful products because of the larger design issues they raise. Technology is simply a tool to serve people, and, obviously, people live in significantly different societies and cultures. We all know that different cultures adapt technologies in different ways to suit their local preoccupations and concerns. And I have certainly had my own localized concerns these past weeks.

Even when the technical fixes are easy, there are design dilemmas and economic trade-offs that arise. True, Macromedia could implement better JPEG rendering in Flash, but that comes at a certain expense: A good renderer is a hard thing to write (even if they could license the photo-rendering code from the Mozilla folks). Also, what they have appears to be good enough for most people (except for me obviously). So, when do you decide that your product is good enough that you can stop pandering to the Long Tail? Can you *ever* afford to do that? Aren't you in danger of missing out on a vast market opportunity?

Yahoo Photos could certainly implement a better photo-resizing algorithm—although presumably there's a performance penalty to be paid if you use a more color-accurate algorithm (or perhaps a larger resultant image size). Since the Yahoo service operates with tens of millions of users and photos, this could potentially limit the scalability of their platform in a serious way. Conversely, if all Yahoo users switched en masse to Flickr, which uses more expensive algorithms, would their platform be able to handle it? Or would it generate a case of teething problems and turn users off because of poor response times, and so on?

Flickr could very easily provide a JavaScript and native HTML browser screen show alternative to their Flash-driven version (as all the other photo services do). The fundamental reason for a screen show is to display a set of images in sequence. It takes perhaps ten lines of JavaScript code to implement something that will work in virtually every browser that exists. The only benefit of Flash is to provide transition effects between images—to be sure, transitions are flashy and liven up screen shows, but that is a matter of style rather than substance. A major downside of Flash is that it isn't included in a default installation of most browsers; one has to go and actively download and install it. The endemic problems users encounter with the installation of software on local machines are ironically part of the reason many have moved to using the simplified interfaces of the browser and the content of the Web. Flash has historically also been problematic when it comes to accessibility, which is the term that software developers use to describe a program's usefulness to people with disabilities. The use of Flash in the browser raises issues with keyboard navigation or high contrast schemes, which can cause difficulties for the sight or hearing impaired. In short, browser content that is rendered in Flash is opaque to many classes of users, making me wonder how many people they have turned away by not providing a native browser screen show. Flickr uses Flash extensively in the rest of their product, and, in the case of the screen show, its usage serves to discourage people from downloading images. Is this pseudo-digital rights management (DRM) an essential feature of their service? A photo-sharing site that makes it difficult to share photos doesn't sound right to my ears. A possible alternative would be to keep Flash but offer differently colored backgrounds for screen shows, in order to avoid the kinds of contrast issues I encountered. But where would that option show up in the user interface?

Similarly, Photoshop could implement a slight variant of their various Quick Fix and Auto Correct features that would be more attuned to my kind of skin color (indeed I assume that photographers in Africa have written macros or filters that do such a thing). How best then to phrase a global preference in an options dialogue in Photoshop? "Adjust for darker skin tones"? Documentation writers would have a field day finding the right verbiage for such an option. Also what about usability? If you add all these preferences to your product what would your user interface look like? Try typing the "about:config" URL in a Mozilla browser to get a sense of the complexity that modern software developers face.

If there were a huge market for these products and services in Africa (which is unlikely given the low Internet penetration rates and presumably widespread instances of software piracy), the issues I faced would of course be a real problem for the companies in question. There would be demand not just for local language versions (say, a Swahili language version in Kenya) but also for tweaks that would make these services more closely attuned to the prevailing culture and, in this case, ethnic backgrounds.

Over the past 150 years, as photography has evolved into the digital realm, photographers in Africa have had to deal with brighter sunshine and higher contrast, as well as darker skin tones, when processing their photos. The people who install photo laboratory hardware in Ghana, where I come from, always have to recalibrate their equipment to deal with the kind of skin tones that dominate the local market. The factory defaults simply won't do. I've had better results developing film in Ghana than in the United States because I often forget to tell the labs here that they should "watch for skin tones."

I'd expect then that software that was truly local (by which I mean, sensitive to local concerns) might sometimes need not just run-of-the-mill language changes, or even writing system changes, but also, as seems to be the case here, algorithmic adaptations.

As software designers, we try to engineer simplicity and refrain from overwhelming users in their interaction with our services and products. Our main focus is usability—for the individual users, for the business community, and for society as a whole. Yet there are very real and often competing concerns about the application of technology in different cultures. So the next time you see a vaguely worded so-called Turkish option somewhere in your application's configuration dialogues, know that someone somewhere was likely adapting their product for a local market. Join me, though, in saluting the developers, testers, product managers, and designers who collectively worked together to come to that solution. I'd hazard that the tweaking of the product was to fix a deal breaker in some market.

Finally, and for what it's worth, I find endlessly fascinating this notion that cultural sensitivity in technology sometimes necessitates algorithmic adaptation. Maybe though, iterative adaptation in response to local environments—evolution, in short—is the name of the game. Perhaps that's simply the way things should be.

POSTSCRIPT (A YEAR LATER)

I reported my issues to Flickr and later prodded them with some other folktales about their excessive use of Flash. I know I wasn't the only one with complaints, but I'd like to think that my slightly different framing of the issue helped tip the balance: a month later, they stopped using Flash to display images and moved to a much lighter weight HTML user interface. They continue to use Flash to drive screen shows and organize albums, but, to their credit, they expose enough of their internals to allow third parties to step into the breach to provide alternate interfaces if needed. There is still work to do to "fix" Flash rendering of images, but that doesn't concern me much since I don't use it. Adobe Photoshop will be a challenge however. I'll simply note that about 5 to 10 searchers now come across this article every day trying to solve the mystery of "darker skin tones Adobe Photoshop," pointing to an unfulfilled need. If that number continues to rise, I'll hazard it won't be long before we see a "dark skin tone" option appearing in the Photoshop preferences. I look forward to that day.

Justin Mullins

Ups and Downs of Jetpacks

Daredevil jetpack travel has never really caught on with commuters, and for good reason. But is all that about to change?

By day Stuart Ross is an airline pilot. By night he dreams of flying. The sort of flight he has in mind is a million miles from his daily routine: no air traffic control, no passengers, and definitely no wings. He dreams of leaping into the air from a standing start, jumping clear over his house, halting motionless hundreds of meters up to admire the view, and then descending gracefully back onto his lawn.

Ross is tantalizingly close to that goal. He has spent two years and £50,000 building his own shiny jetpack. Along the way, he has scorched his clothes and garden with a fuel so unstable that in a recent accident it turned a section of the United Kingdom's busiest motorway into a blazing inferno. And he has almost broken his neck. But at last he's nearly ready to step outside and take to the air.

The first jetpack flew more than 40 years ago, so you might expect that by now designs would be impressively slick—perhaps even ready for daredevil commuters who want to feel the wind on their cheeks. Yet you still can't buy or even rent one. Those that exist remain firmly in the hands of a few diehards like Ross, one of a rare breed of self-taught engineers struggling to overcome the technology's inherent limitations and immense dangers. Their story is itself a white knuckle ride, a tale of lethal rocket fuel, technological bravado, and death-defying bravery but also of greed, kidnapping, and even murder. So what has gone wrong? And will the jetpack ever live up to its promise?

The story begins in the late 1940s with Wendell Moore, chief engineer at the Bell Aircraft Company. Moore was given the task of designing small thrusters to help control the Bell X-1 rocket plane, the first craft to smash the sound barrier. He decided that the best fuel for these thrusters was hydrogen peroxide—a powerful oxidizing agent and bleach. Peroxide decomposes into oxygen and steam when it comes into contact with certain metal catalysts such as silver or copper. Liquid peroxide is easier to handle than most rocket fuels, and the exhaust gases emerge at a few hundred degrees centigrade rather than the thousands that other fuels generate. As a low-thrust fuel, it's hard to beat.

It didn't take long for Moore to think of strapping his thruster onto the back of a man and using it to lift him off the ground. When he patented the idea in the early 1960s, the U.S. army became interested and gave him $150,000 to build a prototype.

Moore's design looks uncannily like the fictional jetpacks used by Buck Rogers and the Jetsons. However, his device is strictly speaking a rocketpack, powered by a peroxide rocket engine rather than a jet. The pilots who flew it during its early tests called the device a rocketbelt, and the name stuck.

The pilot carries the peroxide fuel in two steel tanks mounted on a backpack. A third tank holds pressurized nitrogen that forces the peroxide through a stack of silver mesh disks. The silver catalyzes the decomposition of peroxide into steam and oxygen at 600°C, producing a more than 600-fold increase in volume. This rapid expansion forces the gases through two steel hoses aimed toward the ground at a slight angle behind the pilot's back. The expanding gases accelerate downward and force the machine off the ground, carrying the pilot into the air.

Initial demonstrations stunned the U.S. army. The rocketbelt could reach speeds of up to 100 kilometers per hour in seconds, just the thing to leapfrog soldiers from one spot on the battlefield to another. In its contract with Bell, the army even stipulated that an average GI should be able to fly a rocketbelt with minimal training. So Moore asked Bill Suitor, a friend's teenage son, whether he wanted to try it out. He jumped at the chance, and Suitor eventually became the most famous rocketbelt pilot in the world. He flew the device for American presidents, doubled for Sean Connery in an escape scene in the James Bond movie *Thunderball,* and performed at the opening ceremony of the 1984 Olympics.

Wherever he flew, Suitor won rapturous applause. But these stunt flights glossed over the rocketbelt's one serious disadvantage. The flight at the 1984 Olympics was brief, not because of TV scheduling or modesty on Suitor's part, but because even with full tanks, the rocketbelt can fly for only 25 seconds. Its optimum fuel load is very limited, says Mark Wells, an engineer and rocketbelt expert at NASA's Marshall Space Flight Center in Alabama. "Adding more gives increasingly diminished returns since you must use fuel to lift the extra fuel you carry."

Not only do rocketbelts have limited fuel capacity, but they use this fuel inefficiently compared with a propeller engine, says Wells. A propeller works by accelerating a large amount of air to relatively low speed—a few hundred kilometers per hour—whereas a rocket accelerates a smaller amount of gas to several thousand kilometers per hour. A rocket engine is most efficient at converting its fuel into thrust when moving close to the speed of its exhaust velocity—when the relative velocity of the exhaust and the atmosphere is a minimum. But a rocketbelt is used to simply hover or move relatively slowly, so this is a very inefficient way to fly.

Worse, hydrogen peroxide is a "monopropellant"—it isn't burned or combined with anything else—so pilots must carry all the fuel they need on their back. A jet engine, on the other hand, combines a combustible fuel with oxygen from the atmosphere, which does not have to be carried on board.

So, back in the early 1960s Moore embarked on a project to replace it with a more efficient jet engine, which compresses air between rotating turbines and burns it with kerosene. In 1965, he received $3 million from the U.S. Advanced Research Projects Agency to build a genuine jetpack. His calculations suggested that it should be able to fly at well over 100 kilometers per hour for up to 25 minutes. In 1969, the researchers flew the jetpack for the first time. Its speed and endurance impressed ARPA, but soon after, Moore died suddenly of a heart attack. The project stalled and never recovered.

Over the next 30 years, rocketbelt research continued in a limited way. Copies of Moore's rocketbelt made numerous appearances at major public events, but nobody could find a way round the fundamental 25-second limit, and this killed off almost all military and commercial interest.

The next big hope for personal flight came in 1999, when a company called Millennium Jet based in Santa Clara, California, began building a flying machine called the

SoloTrek. Instead of relying on a rocket or jet engine, the SoloTrek used a piston engine to drive a propeller above the pilot's head. According to company president Michael Moshier, SoloTrek was capable of hovering for up to three hours, flying at 100 kilometers per hour, and traveling more than 200 kilometers.

SoloTrek could hardly be called a jetpack, as the pilot was strapped into a kind of exoskeleton that took the weight of the engine and propeller while the machine was on the ground. But it achieved the same aim. In numerous flight trials at NASA's Ames Research Center in California, the machine appeared to perform perfectly. Then disaster struck.

For safety reasons, the machine was suspended on a retracting tether during test flights. The tether system was designed to automatically reel in as the SoloTrek rose from the ground, so that if the device lost power it could be lowered gently instead of plummeting. But during a flight shortly after a rain shower in 2002, the tether failed to retract and tangled in the propeller blades, which then disintegrated. The machine and pilot dropped to the ground, and while the pilot walked away unharmed, the vehicle was damaged beyond repair. Unable to stick to the tight development schedule that its backers demanded, Millennium Jet lost its funding, and in 2003 it closed down.

You might think that the successive failures of the rocketbelt, jetpack, and SoloTrek would have killed off the dream of simply strapping on an engine and leaping into the sky. Yet a small number of enthusiasts are keeping that dream alive. Part of the reason is that with just a couple of working rocketbelts worldwide, they can command upward of $20,000 per flight for film and publicity appearances. That kind of money focuses the mind. Unfortunately it doesn't always bring out the best in people.

KIDNAP AND MURDER

In 1992, onetime insurance salesman and entrepreneur Brad Barker formed a company to build a rocketbelt with two partners: Joe Wright, a businessman based in Houston, and Larry Stanley, an engineer who owned an oil well in Texas. By 1994, they had a working prototype that they called the Rocketbelt-2000 or RB-2000. They even asked Suitor to fly it for them.

But the partnership soon broke down. First Stanley accused Barker of defrauding the company. Then Barker attacked Stanley and went into hiding, taking the RB-2000 with him. In July 1998, Wright was found beaten to death at his home. Police investigators questioned Barker but released him after three days. The following year Stanley took Barker to court to recover lost earnings. The judge awarded Stanley sole ownership of the RB-2000 and over $10 million in costs and damages. But when Barker refused to cough up, Stanley kidnapped him, tied him up, and held him captive in a box. After eight days Barker managed to escape. Police then arrested Stanley, and in 2002 he was sentenced to life in prison, since reduced to eight years. The rocketbelt has never been found.

The story of the RB-2000 has not deterred Stuart Ross, however. Two years ago, Ross, who lives in England in a peaceful Sussex village, began to build his own rocketbelt using photographs of existing designs and some equipment and plans bought from a failed rocketbelt builder in the United States. At the same time, Ross also had to learn to handle the peroxide rocket fuel.

Industrial-strength peroxide has a concentration of roughly 60 percent. To work as rocket fuel, this must be distilled to reach 87 to 90 percent. At this strength, it is hugely reactive: Spill 90 percent peroxide onto a piece of wood and it can burst into flames. If it touches copper or silver, it decomposes instantly into hot steam and oxygen gas, which can react explosively if it comes into contact with anything flammable.

To help prevent such accidents, peroxide manufacturers add stabilizers to their product. The identity of these stabilizers is a commercial secret, but they work by bonding to the surface of metals to prevent the peroxide decomposing. This is bad news for people like Ross: the stabilizers stop the silver catalyzing the fuel's decomposition, making it useless. In one incident 20 years ago, a rocketbelt pilot who was using peroxide claimed that his machine failed in midflight because traces of stabilizer had somehow contaminated the rocketbelt engine. The pilot then successfully sued his fuel supplier, with the result that manufacturers now refuse to supply concentrated peroxide to individuals.

Fortunately for Ross and other rocketbelt flyers, they can turn to Eric Bengtsson, a chemical engineer based in Sweden. Bengtsson specializes in making peroxide and has developed a stabilizer that does not contaminate the silver catalysts used in rocketbelts. "I know how this class of chemical works and that some stabilizers are less problematic than others. I eventually came across one that stabilized the peroxide without contaminating silver," he says. He removes the manufacturer's stabilizer and replaces the stuff with his own and is happy to supply enthusiasts with fuel or even help them set up their own distillery.

The risk of accident due to spillage remains. Just over a month ago, for example, a delivery of concentrated hydrogen peroxide fuel on its way to Ross sprang a leak on one of the United Kingdom's busiest motorways, the M25. The truck that was carrying it—and a large section of the road—burst into flames, shutting the motorway for hours and causing chaos, though there were no serious injuries.

Ross finally completed his rocketbelt earlier this year, and on April 20 he attempted his first takeoff. Just in case anything went wrong he tethered the rocketbelt to a safety cable, but the device performed perfectly. Over the next few months he flew a total of 12 tethered flights, and all went well. Then in August, on his 13th flight, his luck ran out.

Shortly after Ross lifted off outside his home, the rocketbelt's throttle jammed open. Attached to the ground by a cable, Ross was flung from side to side like a deflating balloon. Eventually he was able to cut the fuel supply, but the incident could easily have killed him. "I was worried that the steel tether would wrap around my neck," he says.

The problem stemmed from the huge pressures needed to make a rocketbelt fly. To pump the peroxide into the decomposition chamber, the liquid is pressurized to about 60 atmospheres. The rocketbelt can only be throttled by controlling this flow, but small changes in the flow can have a dramatic impact on thrust. The most difficult challenge that Ross and other rocketbelt builders face is to construct a throttle that can safely control the amount of peroxide passing into the decomposition chamber. Numerous enthusiasts have fallen at this hurdle, and Ross's accident showed him how badly things can go wrong.

It is no trivial engineering challenge. Fully fueled, Ross and his rocketbelt weigh about 133 kilograms. To create enough thrust to exactly balance this weight, the throttle must allow precisely 0.87 liters of peroxide per second into the decomposition chamber. The maximum safe flow rate, however, is just 0.91 liters per second, which creates 150 kilograms of thrust, enough to accelerate the pilot to substantial speeds. The throttle has to be able to smoothly and reliably control this tiny change. "I know of no other application where the tolerances are so tight," says Ross.

The throttle he used on his August flight consisted of a

piston in a tube that has a row of inlet holes on one side and an exit port on the other. Withdrawing the piston uncovers the holes one after another, increasing the flow of peroxide in small increments.

Ross tested the design in numerous static runs with the rocketbelt anchored to the ground. The flow through the throttle at these pressures is hugely complex, and the tests were designed to find out whether it was reliable under all the conditions the device would be likely to experience during flight. Everything seemed fine.

However, during his fateful flight Ross opened the throttle too quickly. The sudden increase in pressure pushed the piston to its fully open position and held it there, a situation known as hydraulic lock. Ross hadn't realized his throttle could jam like that. "My next flight was going to be untethered," he says. "There's enough fuel to reach 8,000 feet if you fly straight up."

Ross approached Matt Linfield, an engineer who runs Linfield Precision Engineering near Ross's home, to come up with a better throttle. Linfield chose a design based on a needle valve, commonly used in the automotive and aerospace industry, in which a long, tapered needle sits inside a similarly shaped sleeve. The valve is opened by withdrawing the piston, thus opening up a gap between the needle and the sleeve and allowing fuel to flow through. Pushing the piston back into the sleeve closes the valve. The valve is fail safe: fuel flows in at the wider end of the sleeve and out from the narrow end, so it exerts pressure on the piston that ensures the default position is shut. "You always want it to be fail safe," says Linfield.

Ross is now putting the new design through its paces in static tests. So far everything looks good, and Ross believes he will be airborne before the end of the year. Eventually he plans to fly the device for a fee at high-profile events around the world, just as Suitor did in 1984.

But what about the rest of us? Ross has no illusions about the future of rocketbelts. Their limited flight time and inherent danger mean that few people will ever get to fly one. Ross has little doubt that this can only be a good thing. "Can you imagine if everyone had one of these?" he says. "It would be chaos."

Jesse Sunenblick

Into the Great Wide Open

New technology could radically transform broadcasting. The dreamers and players are already debating how far to go.

In 1940, the Austrian-born actress Hedy Lamarr, considered by some the most beautiful woman in Hollywood, approached her neighbor there, the avant-garde composer George Antheil, and asked him a question about glands. Antheil, known for his propulsive film scores for multiple player pianos, had broad interests: In addition to his music he wrote a syndicated advice-to-the-lovelorn column and had even published a medical book, *Every Man His Own Detective: A Study of Glandular Endocrinology*. As the story goes, Lamarr—whose acting exploits (which include the first big-screen nude scene) and marriages (there were six husbands, most notably Fritz Mandl, an Austrian arms dealer with ties to Hitler and Mussolini) are too varied to discuss here except to say that she was a woman far ahead of her time—wanted to know how she might enlarge her breasts. Somehow, though, they ended up talking about radio-controlled torpedoes, and the future of communications was changed.

After years of living with Mandl, Lamarr was familiar

with the problem of sending control signals to a torpedo after it was launched from a ship, especially radio signals, which the enemy could easily detect and jam. She had a notion of a radio transmission that, by changing its frequency many times a second, could allow an observation plane to covertly guide a torpedo over long distances. Combining Lamarr's knowledge of radio control with the model Antheil had used to coordinate 16 pianos in his *BalletMécanique,* the pair invented the idea of "frequency hopping" and obtained a patent for a Secret Communications System. This was the first example of a single radio transmission using multiple frequencies across the radio spectrum—the range of electromagnetic frequencies that are useful for sending broadcast signals—without bumping into other transmissions and causing interference. Sixty-plus years later, frequency hopping has evolved into a technology, called "spread spectrum," that proponents claim could put an end to most forms of radio interference, presaging a time when the airwaves (TV signals travel over the same spectrum), one of our most heavily regulated resources, could be opened up.

The implications of this idea are far reaching for human communication, including journalism. If there is no longer a reason to tightly regulate the broadcast spectrum, then just about anyone would be allowed to broadcast. As technology continues its march toward miniaturization and higher speeds, we might soon have devices that fit in our pockets capable of sending voice, video, and other data over long distances. And if we could use such devices without causing interference, then today's bloggers, for example, confined by laptops, short-range wireless connections, and slow video feeds, could be tomorrow's roving band of telejournalists. Imagine lone-wolf Christiane Amanpours showing up on site, unencumbered by the demands and the strictures of our modern media monopolies, beaming reports live to whoever might care to watch, not just on television but on a computer, on a cell phone, on the dashboard of a car.

As in the earliest days, broadcast pioneers are once again talking and dreaming about broadcast's potential to connect all corners of the earth. Of course, in the world of broadcasting what is possible is often undone by what is profitable—or politically expedient. The advent of spread spectrum has spawned a subterranean debate about how to manage the radio spectrum that has broadcasters arguing with technologists, economists arguing with media critics, and everybody arguing with the FCC about a radio revolution.

When you connect to the Internet at Starbucks, when you talk on your cell phone, or when you use many of the other radio technologies that constitute our current wireless craze, you are using spread spectrum. Spread spectrum works by contradicting the traditional rules of radio communication, in which a single signal is sent over a single frequency in the electromagnetic spectrum for which it has a license from the FCC. With spread spectrum, a transmission is disassembled and sent out over a variety of frequencies, without causing interference to whatever else might be operating within those frequencies, and is reassembled on the other end by a "smart" receiver. Licenses aren't necessary for spread-spectrum transmissions, but the devices currently aren't allowed to operate at more than a few watts of power. And since the early 1990s, when they were first available for use by consumers, they have been relegated to that portion of the radio spectrum known as the "junk band"—the uppermost usable frequencies that are home to gadgets like cordless phones, microwave ovens, and baby monitors and which, because of shorter wavelengths, have trouble cutting through bad weather and obstacles like trees and buildings.

The FCC has issued licenses for frequencies since it was established in 1927, and the impetus to do so was an outgrowth of a decade of ethereal chaos in the 1920s, when the airwaves were overloaded with so many new broadcasters on so few available frequencies that it was impossible in many urban areas to receive a steady signal. Media critics like to point out how this licensing system has contributed to an oligarchy of the air, in which the Viacoms and Clear Channels of the world control access to most radio communication. But by 1990 it had contributed to something else: a dearth of available frequencies left to license. The spectrum, like an oil reserve, was nearly depleted.

Spread spectrum offers a far more efficient way of using the radio spectrum, and throughout the 1990s the FCC opened up license-free slivers for devices that employ spread-spectrum technology—first for gadgets like garage door openers and home alarm systems and later for WiFi, which has blossomed into a multibillion-dollar industry. WiFi not only allows city dwellers to hook up to the Internet at Starbucks but is pushing the Internet into rural locales not served by cable or DSL and making possible public-safety networks for police and fire departments.

Now the FCC is considering a series of rule changes that would open up much more of the spectrum for unlicensed radio. The timetable on any commission decision on such rule changes is fluid and depends, in part, on who replaces Michael Powell, a strong proponent of unlicensed radio technology, as FCC chairman. The most significant of the rule changes would allow unlicensed radio to operate with more power, over longer distances, and in portions of the spectrum currently occupied by heavyweight incumbents such as the television networks; they would also clear the path for the manufacture of smart radios, which can transmit selectively through little-trafficked frequencies, essentially dodging interference. The big broadcasters are engaged in a rigorous lobbying effort to discredit the science of spread spectrum, which they believe could undercut their competitive edge by allowing thousands of individuals to establish their own television or radio programming or to offer wireless Internet service on the cheap. To public-interest groups, however, the advent of unlicensed radio represents an opportunity for greater citizen access to the airwaves and the possibility of a network of community radio or TV stations in every town in America.

"The rule changes represent the most important communications decision the FCC will face in the next 10 years," says Harold Feld, associate director of the Media Access Project, a nonprofit, public-interest telecommunications law firm that hopes the FCC will expand the role of unlicensed radio. "If the commission can stand up to the most powerful industry lobbies in Washington and create new rules that reflect new technologies, the American people will see nothing short of miracles."

In my conversations with Feld, he kept repeating the phrase "cheap, ubiquitous Internet access"—which, in his opinion, is the crux of the debate—and emphasizing the importance of getting these spread-spectrum devices deployed with sufficient power and with access to the lower frequency, "beachfront" sections of the radio spectrum dominated by the big broadcasters. That, he says, would create a plethora of journalistic opportunities for media big and small. In addition to creating a nation of broadcasters, network news companies could bolster their Web offerings with live-action video feeds, using a one-person news crew, from anywhere with a WiFi connection. A WiFi media reader, meanwhile, could replace the bundle of newspapers and magazines that you carry to work or home every day. And ubiquitous mobile Internet connections would mean that reporters, who would have constant access to research tools, could improve the content of their stories. (Of course, someone would have to pay for all these technological goodies.)

Everyone, it seems, has a dog in this fight. Venture capitalists who stand to make a buck off more powerful versions of WiFi. Technologists who want an arena for their futuristic ideas. Media activists who envision an unlimited radio dial with thousands, if not millions, of noncommercial stations. Wireless Internet service providers who want to extend their reach. Economists who think that the best way to make use of new radio technology is to privatize the radio spectrum and let the instincts of capitalism take over. And of course lobbyists for powerful incumbents who want to preserve their exclusive licenses.

Out of this fray has come a distinctive vision of the spectrum as a public commons, in which an unlimited number of users share unlicensed portions of the radio spectrum, and—subject, of course, to power and usage restrictions—do with it what they want. The movement gained steam throughout the 1990s, as advancing spread-spectrum technologies called the FCC's licensing system into question. One of the movement's philosophical pillars is that unlicensed radio technology has the ability to democratize the media, much the same way that the Internet did through blogging, although on a profoundly grander scale. Eben Moglen, a Columbia University law professor and one of open spectrum's biggest supporters, has an idea about how open spectrum might accomplish this.

For the last 11 years, Moglen has served as general counsel (pro bono, of course) to the Free Software Foundation, a group that promotes the creation and distribution of, well, free software. He is also an unabashed Marxist. In his office, I told Moglen I was having trouble understanding how an "open" spectrum would differ from a "closed" spectrum and could he please offer an analogy from the real world that would bring the otherworldliness of the radio spectrum into context.

He leaned back in his chair and spread his arms out wide, as though everything around us were part of the analogy he was about to give. Which was true. "Take the island of Manhattan," Moglen said. "The level of anonymity in Manhattan is subject to social regulation, like the radio spectrum is subject to political regulation. And it's variable: Sometimes you go places where you have to identify yourself, sometimes not. And as the city imposes restrictions on movement, zoning, and behavior, the federal government places restrictions upon the radio spectrum.

"However," he went on, "the difference is in the number of restrictions. The essence of life in Manhattan is openness. It's all free, it's all here, you can get to it. You can walk from the West Side to the East Side, from Harlem to Chinatown. Or take a cab. But what would Manhattan look like if its social policies were on par with the current government policies concerning use of the radio spectrum? It would be unendurable. You'd have David Rockefeller owning Rockefeller Center. Rupert Murdoch would have a dominant say in everything that happens between the Battery and 23rd Street. Worse, you'd be sitting in Starbucks, having a conversation, and somebody would say, 'You, stop talking! You, talk about the weather!' You're allowed to have person-to-person conversations, but for the privilege of doing so in my neighborhood you have to pay six dollars a minute. And, because there's nothing resembling a Central Park on the radio spectrum, if you want to gather people and talk about the war in Iraq, tough luck! We need a Central Park for radio!"

If the government tried to license newspapers, Moglen says, the courts would block it on the ground that it violated the First Amendment. "The technological reason that we have given in the past for why a system of licensing—one that would be completely unconstitutional with respect to print—is constitutional in the spectrum no longer exists! And when the broadcasting licensing system falls, as it inevitably must, American society will be transformed. Mr. Murdoch, Mr. Eisner, Mr. Gates—Will. Be. Poorer. We. Will. Be. Richer. And there will again be news in this society, which at the moment, there almost isn't."

Many media critics accuse the FCC, under Chairman Powell, of perpetuating communications policies that favor forces of media consolidation and the status quo. Yet interestingly, in the case of unlicensed wireless communication, Powell has been on the other side of the barricades from big media. To Powell, WiFi is the prototype for the role that unlicensed radio will play in the future, an example of what he has described, in various speeches, as "the great Digital Migration" or "the Age of Personal Communications"—optimistic assessments of what he sees as a new information paradigm that lies just over the horizon. In terms of convenience, at least, we are certainly on the cusp of profound changes: We will soon talk over the Internet the way we talk today over telephones, but for less money, because Internet voice is a computer application, not a government-regulated telecom, and because providers don't need to build a multi-billion-dollar infrastructure to offer it. This is likely to give rise to an Internet of things, a state of überconnectivity. "The visionary sermons of technology futurists seem to have materialized," Powell said in a January 2004 speech at the National Press Club in Washington, an assessment of the transition from the world of analog to digital. "No longer the stuff of science fiction novels, crystal balls, and academic conferences, it is real. Technology is bringing more power to the people."

But a running theme in Powell's Washington speech is that the "Digital Migration" means far more to Americans than convenience—that the ubiquity of the Internet, combined with the miniaturization and higher power of radio technology, will empower individuals, rather than large institutions, to become central in the creation and dissemination of ideas. "Governments are almost always about geography, jurisdiction, and centralized control," Powell said in his Washington speech. "The Internet is unhindered by geography, dismissive of jurisdiction, and decentralizes control." The implication here is that technology, if given the chance, will level the playing field. Toward the end of his Washington speech the chairman laid out a multipronged strategy for accomplishing the "migration," much of it dependent on reforming the radio spectrum to allow the next generation of unlicensed radio to operate with more power and bandwidth—so that a wireless Internet network that can now reach a distance of 200 feet, for example, might some day spread across 200 miles or perhaps the entire planet. Whoever gets Powell's job is unlikely to share his passion for unlicensed radio, but the debate isn't going away.

Dave Hughes, a retired colonel from Colorado Springs, knows better than most what a network like the one Powell described might look like. It was Hughes who told me the story of Hedy Lamarr, because it was Hughes who had resurrected her name from the dustbin of technological afterthoughts when he nominated her for an achievement award (which she ultimately won) from the Electronic Frontier Foundation, a watchdog for digital civil liberties, in 1997, when Lamarr, who died in 2000, was old and forgotten and living in Florida. ("It's about time," Lamarr is rumored to have said upon hearing of the award.) Colonel Hughes has been a lot of things in life—a hero of the wars in Korea and Vietnam, a professor of English at West Point, an inventor of America's first online computer bulletin board, and a pioneer of rural computer networking—but what he is most is a vociferous advocate of radio and in particular of spread spectrum.

"My humble goal," Hughes likes to say, "is to see all 6 billion minds on the planet connected in all the ways our brains and ears and mouths and eyes can communicate. At least when you can communicate, you can reduce the areas of disagreement to real substance."

In 1991, Hughes bought two of the earliest spread-spectrum radios to hit the marketplace, units that produced a single watt of power. He connected them between his Internet-equipped office building and an early IBM version of a Web site on his home computer. The wireless link of that connection between the two radios cost him nothing; had U.S. West provided the connection, it would have cost $600 a month. Hughes thought: If it works between buildings, why not rural towns? In the mountains of southwest Colorado, Hughes perched a pair of spread-spectrum radios in such a way that a school district in the town of San Luis was connected wirelessly to an early version of the Internet, for a one-time cost of $3,000, as opposed to the $2,000 a month that U.S. West was asking at the time to run a 40-mile cable to Alamosa, where the closest Internet provider was. Then he thought, if we can connect towns here, why not in the most remote places on Earth? He went to Mongolia with spread-spectrum radios, and now Ulan Bator is the third world's most wirelessly connected city.

In 2003, Hughes used three WiFi radios in his most ambitious project yet: constructing the world's highest

Internet café at the base camp of Mount Everest. From his home in Colorado, Hughes collaborated with Tsering Gyaltsen, the grandson of the only surviving Sherpa to have accompanied Sir Edmund Hillary on his first ascent of the mountain, and designed a network in which WiFi radios in the Café Tent beneath the Khumbu Ice Fall are linked to a satellite dish 1,500 feet away that sends data, via satellite, to an Internet service provider in Israel. Then, last year, Hughes helped another Nepalese entrepreneur add computers and a wireless link to his cybercafé in the Namche Bazaar, the trading center of the Everest region. He used three antenna relays (one hanging off the side of a monastery at 14,000 feet) to extend the network to a school in the nearby town of Thame, where 10 Sherpa children are now taking English and computer classes over the Internet from a Nepalese-born, English-speaking Sherpa computer programmer who lives in Pittsburgh.

The computers the Nepalese children use rely on a free software program called Free World Dial Up, which allows them to speak to their teacher over computers, for a flat rate, the way most people do over telephones. To Hughes, when Nepalese children are talking to their teacher in Pittsburgh and handing in their lessons by computer, giving them, as he says, "half a fighting chance to succeed in this world," that represents more than a demonstration of the possibilities of technology. It is a paradigm shift, a revolution. "You have to understand the disruptive nature of this technology," he often says. "Who's getting robbed? Because of the technology, it's AT&T that's getting robbed. The technology is way out in front of the regulatory, legal, and economic communications systems of this country. There's going to be titanic battles. But ask yourself, What happens to the incumbents? Well, what happened to the horse and buggy? What happened to the printing press?" Of course, the revolution

Hughes envisions can always be interrupted by the real world. On February 1, Nepal's king Gyanendra dissolved the government and shut down all telephone and Internet connections in the country. A WiFi network—even a global one—could not stop a power grab in Nepal.

It is easy to be nostalgic for the earliest days of radio, when, before a licensing regime was put in place, tens of thousands of amateur operators shared the still-mysterious airwaves in a raw, often free-form haze of chatter, music, and news, much like the Internet today. There was something supernatural about radio then, and one line of thinking saw the medium as a force of social connectivity. An article from *Collier's Magazine* in 1922 entitled "Radio Dreams That Can Come True" talks about radio "spreading mutual understanding to all the sections of the country, unifying our thoughts, ideas and purposes, making us a strong and well-knit people." There is an obvious utopian quality to such declarations, common in all periods of significant technological change. While one could certainly argue that first radio, and then television, did achieve the task of uniting us, it did not happen in a way that the *Collier's* writer could have imagined, and it did not happen in a way that had all that much to do with making us strong and well knit.

By the mid-1920s radio was controlled by two national networks—NBC and CBS—and the coming years would witness the invention of advertising and the fine-tuning of capitalism on the radio dial. Does the same fate await spread spectrum and smart radios? In the mainstream press, technoenthusiast feature stories appear regularly, touting the cutting edge of unlicensed wireless communications like WiMax, which is essentially a pumped-up version of WiFi, with a networking reach of 30 miles; Zigbee, a tiny wireless sensor that can be placed on crops to track heat, moisture, and nutrients in the soil; and Ultrawideband, an emerging technology that can move huge amounts of data over short distances. The attendant prophecies sound transformative. "These technologies will usher in a new era for the wireless Web," *Business Week* declared last April. "They'll work with each other and with traditional telephone networks to let people and machines communicate like never before." Lost in such assumptions is a legitimate chance that even if unlicensed devices like smart radios become available to the public, regulators would compromise the potential of such equipment by, for example, imposing strict power limitations to avoid even the slightest chance of interference.

And then there are the economic realities. One of the groups fighting the hardest to open up the radio spectrum to unlicensed radios is the New America Foundation, a public-policy think tank in Washington. On its Web site, New America offers a minitreatise on solutions to our current spectrum woes and comes across as a voice of the people, advocating, among other things, "greater shared citizen access to the airwaves." Yet unlike many supporters of unlicensed radio, New America's vision contains a seed of economic prudence. Jim Snider, a research fellow there who specializes in spectrum issues, was quick to point out not only the "open" and "unmediated" nature of unlicensed radio but also the rising tide of interest from the venture capital community, which over the last two years has produced over 20 well-financed startups and a variety of new products.

Snider advised me to visit the offices of Shared Spectrum Company, a venture capital firm in Virginia that builds prototype frequency-agile radio transmitters (which hop from channel to channel across wide swaths of the radio spectrum, looking for quiet places to transmit) and whose efforts New America endorses. And so, on a stultifying day in late July, I took a ride on the Metro out to Tyson's Corner, Virginia, the East Coast's consummate "edge city"—those hybrid constellations of retail, office, and residential developments near highway interchanges and an older, central city that are paragons of the new economy.

Shared Spectrum rented an office on the second floor of a building with tinted windows, across the hall from a travel agency called Vacation Station. When I arrived, its founder, Mark McHenry, accompanied by his lawyer, suggested we go to the roof so that he could show me what a vast, empty wasteland the supposedly crowded spectrum really is. On the roof, a young employee had set up an antenna and an expensive machine called a spectrum analyzer, a boxy device that sweeps through every radio frequency and displays, on a screen, how much signal strength, which McHenry referred to as "energy," is operating at each frequency. The four of us got down on hands and knees to watch the machine work. Zipping through swaths of spectrum, it immediately made clear why McHenry had such confidence in smart radios and why the prospect of building them had enticed him to leave a cushy job as a program manager at the Defense Advanced Research Projects Agency, or DARPA, the furtive technology arm of the Pentagon widely credited with having invented the Internet.

We zoomed in on the aviation band, where there was little activity. Then the TV band, where there were gaps all over the place. The military band, eerily dead. "There's basically nothing here," McHenry said. He beamed. "Once you accept the idea of frequency-agile radios, anything becomes possible."

Back inside, I asked if he (like Michael Powell, Dave Hughes, and Eben Moglen) thought that smart radios would empower people to become active participants in the creation of knowledge. I had assumed that McHenry, like the folks at the New America Foundation, would see these gadgets as an egalitarian force. But his response—a slight shake of the head and a bewildered look—made me feel silly for asking.

Back in New York, I thought of something McHenry had told me about the U.S. military's plans to use spread-spectrum technology in warfare. I remembered his saying the word *robots,* and so I did some research and found what seemed like a good window into the nexus between technology and corporate and military power. DARPA, the outfit McHenry used to work for, is in the process of developing what it calls Next Generation, or XG, technology—the mother of all spectrum-sharing protocols—that will enable every unit on the battlefield to communicate by radio, over longer distances and with more ease of use than is currently possible. Using smart technology, XG radios will store the spectrum conditions for every country on Earth on a microchip and automatically conform to the conditions of the environment, avoiding the hassle of manually assigning frequencies to military radios during combat. Shared Spectrum is getting paid millions of dollars to help DARPA develop algorithms for XG radios. As I clicked further into the bowels of various military Web sites, I came across another organization, the Artificial Intelligence Center, which has a hand in several army projects, including TEAMBOTICA—radio-controlled robots that the army plans to use in reconnaissance and surveillance missions.

As I looked at pictures of the TEAMBOTICA robot on my computer, the words of Mark McHenry echoed in my ears: "Once you accept the idea of frequency-agile radios, anything becomes possible." Indeed. Eben Moglen and Dave Hughes had said essentially the same thing. With the exception of heavyweight spectrum incumbents like the broadcasters, who are unable or unwilling to concede the end of interference, most everyone who talks about unlicensed radio uses the same vocabulary, although to radically different ends. For Moglen it is about democracy. For Hughes it is about connectivity. And for McHenry it is about money. His frequency-agile radios have already entered the military-industrial complex; someday soon, this technology will likely enter the civilian realm and forge a path not unlike the Internet, making a few people very rich, producing devices that we might come to see as indispensable but that in the end may or may not have much to do with freedom, personal or otherwise.

Evan Ratliff

The Zombie Hunters

On the trail of cyberextortionists

One afternoon this spring, a half dozen young computer engineers sat in the headquarters of Prolexic, an Internet-security company in Hollywood, Florida, puzzling over an attack on one of the company's clients, a penile-enhancement business called MensNiche.com. The engineers, gathered in the company's network operations center, or noc, on the fourth floor of a new office building, were monitoring Internet traffic on 50-inch wall-mounted screens. Anna Claiborne, one of the company's senior network engineers, wandered into the noc in jeans and a T-shirt. The MensNiche attacker had launched an assault on the company's Web site at 4 a.m., and Claiborne had spent the night in the office fending it off. "Hence," she said, "I look like hell today."

MensNiche's problems had begun a week earlier, with a flood of fake data requests—what is known as a distributed denial-of-service attack—from computers around the world. Although few, if any, of those computers' owners knew it, their machines had been hijacked by hackers; they had become what programmers call "zombies" and had been set loose on MensNiche. The result was akin to what occurs when callers jam the phone lines during a television contest: With so many computers trying to connect, almost none could get through, and the company was losing business.

The first wave of the attack was easily filtered by Prolexic's automated system. The assailant then disguised his zombies as legitimate Web users, fooling the filters so well that Claiborne refused to tell me how it was done, for fear that others would adopt the same tactic. She spent the night examining the requests one by one as they scrolled by—interrogating each zombie, trying to find a key to the attacker's strategy.

"He's clever, and he's been trying everything," Claiborne said. "If we ever find out who it is, seriously, I'd be willing to buy a plane ticket, fly over, and punch him in the face."

Prolexic, which was founded in 2003 by a 27-year-old college dropout named Barrett Lyon, is a 24-hour, seven-days-a-week operation. An engineer is posted in the noc at all times to monitor Prolexic's four data hubs, which are in Phoenix, Vancouver, Miami, and London. The hubs contain powerful computers designed to absorb the brunt of data floods and are, essentially, massive holding pens for zombies. Any data traveling to Prolexic's clients pass through this hardware. The company, which had revenues of 4 million dollars in its first year, now has more than 80 customers.

Lyon's main business is protecting his clients from cyberextortionists, who demand payments from companies in return for leaving them alone. Although Lyon is based in Florida, the attackers he deals with might be in Kazakhstan or China, and they usually don't work alone.

"It's an insanely stressful job," Claiborne told me. "You are the middleman between people who are losing thousands or millions of dollars and somebody who really wants to make that person lose thousands or millions of dollars." When the monitors' graphs begin to spike, indicating that an attack is under way, she said, "it's like looking at the ocean and seeing a wall of water 300 feet high coming toward you."

Only a few years ago, online malfeasance was largely the province of either technically adept hackers (or "crackers," as ill-intentioned hackers are known), who were in it for the thrill or for bragging rights, or novices (called "script kiddies"), who unleashed viruses as pranks. But as the Web's reach has expanded real-world criminals have discovered its potential. Mobsters and con men, from Africa to Eastern Europe, have gone online. Increasingly, cyberextortionists are tied to gangs that operate in several countries and hide within a labyrinth of anonymous accounts.

"When the attack starts, the ticker starts for that company," Lyon said. "It's a mental game that you've been playing, and if you make a mistake it causes the whole thing to go down. You are terrified."

Lyon, as usual, was wearing shorts and flip-flops. He has blond hair, a trim build, and narrow hazel eyes that were framed by dark circles of fatigue. A poster for the 1983 movie *WarGames*—a major influence—hung above his desk, on which were four computer monitors: one for writing program code, one for watching data traffic, one for surfing the Web, and one for chatting with customers. Lyon leaned over and showed me a program that he had created to identify the zombies attacking MensNiche. When he ran it, a list of countries scrolled up the screen: the United States, China, Cambodia, Haiti, even Iraq.

Examining the list of zombie addresses, Lyon picked one and ran a command called a "traceroute." The program followed the zombie's path from MensNiche back to a computer called NOCC.ior.navy.mil—part of the United States Navy's Network Operations Center for the Indian Ocean Region. "Well, that's great," he said, laughing. Lyon's next traceroute found that another zombie was on the Department of Defense's Military Sealift Command network. The network forces of the U.S. military had been conscripted in an attack on a Web site for penis enlargement.

Michael Alculumbre's first communication from the extortionists arrived on a Thursday evening in August 2004. An e-mail message was sent to him just after 8 p.m. at Protx, an online-payment processing company based in London, where he is the chief executive officer. The subject line read, simply, "Contact us," and the return address—commerce_protection@yahoo.com—offered no clues to the message's origin. The note was cordial and succinct, written in stilted English. "Hello," it began. "We attack your servers for some time. If you want save your business, you should pay 10.000$ bank wire to our bank account. When we receive money, we stop attack immediately. If we will not receive money, we will attack your business 1 month." The note said that $10,000 dollars would buy Protx a year's worth of protection. "Think about how much money you lose, while your servers are down. Thanks John Martino." Alculumbre had never heard of John Martino. He decided to ignore the demand.

Two months later, Alculumbre's network technician called him at home. He said that customers were complaining that the system was off-line. By the time Alculumbre arrived at the office, the source of the disruption was clear. Thousands of computers were inundating Protx's Web site with fake data requests. Many of Protx's legitimate customers received the Internet equivalent of a busy signal—a message saying that the company's servers weren't responding.

Every minute that the Web site remained off-line, Protx's business suffered. As the company's engineers struggled to contain the attack, another $10,000 e-mail demand arrived, this time signed "Tony Martino." Again, Alculumbre ignored it. He had received a call from an agent of the British National Hi-Tech Crime Unit, which had been monitoring the attack. The agent let him know that paying Martino wasn't an option; the extortionist would only return. Beyond that advice, there wasn't much that the NHTCU could do to help. By the time Alculumbre's engineers were able to get the site running, it had been disabled for almost two days.

Alculumbre heard from Tony Martino again the following April, when he received a message offering a $1,000-a-month protection-money payment plan. Before he could respond, an army of up to 70,000 zombies ripped through Protx's defenses and knocked its Web site off-line. This time, it took Protx's engineers three days to fight off the attack.

The company now spends roughly $500,000 a year to protect itself—50 times what Martino had asked for. This includes a $100,000-a-year security contract with Prolexic. Martino, it turned out, had been targeting Lyon's clients for months before he hit Protx.

"This is very similar to the pubs and clubs in London 40 years ago that used to pay money to not have their premises smashed up," Mick Deats, the deputy head of the NHTCU, told me. "It's just a straight, old-fashioned protection racket, with a completely new method." The cyberextortionists also make use of an elaborate money-laundering system, Deats said. "They have companies registered all over the place, passing the money through them."

"I started prosecuting network-attack cases in 1992, and back then it was more the sort of lone hackers," said

Christopher Painter, the deputy chief of the Computer Crime and Intellectual Property Section at the Department of Justice. Today, he says, "you have organized criminal groups that are adopting technical sophistication."

The most potent weapon for Web gangsters is the botnet. A bot, broadly speaking, is a remote-controlled software program that is installed on a computer without the owner's knowledge. Hackers use viruses, worms, or automated programs to scan the Internet in search of potential zombies. One recent study found that a new PC, attached to the Internet without protective software, will on average be infected in about 20 minutes.

In the most common scenario, the bots surreptitiously connect hundreds, or thousands, of zombies to a channel in a chat room. The process is called "herding," and a herd of zombies is called a botnet. The herder then issues orders to the zombies, telling them to send unsolicited e-mail, steal personal information, or launch attacks. Herders also trade, rent, and sell their zombies. "The botnet is the little engine that makes the evil of the Internet work," Chris Morrow, a senior network-security engineer at MCI, said. "It makes spam work. It makes identity fraud work. It makes extortion, in this case, work."

Less than five years ago, experts considered a several-thousand-zombie botnet extraordinary. Lyon now regularly faces botnets of 50,000 zombies or more. According to one study, 15 percent of new zombies are from China. A British Internet-security firm, Clearswift, recently predicted that "botnets will, unless matters change dramatically, proliferate to the point where much of the Internet . . . comes to resemble a mosaic of botnets." Meanwhile, the resources of law enforcement are limited—the NHTCU, for example, has 60 agents handling everything from child pornography to identity theft.

Extortionists often prefer to target online industries, such as pornography and gambling, that occupy a gray area and may be reluctant to seek help from law enforcement. Such businesses account for most of Prolexic's clients. I asked Lyon how he felt about the companies he defended. "Everybody makes a living somehow," he said. "It's not my job to worry about how they do it."

I asked whether that applied to extortionists as well. After a pause, he said, "I guess I'm partial to dot-commers."

Several weeks later, he called me to say that he'd reconsidered his answer. "The Internet is all about connecting things, communicating and sharing information, bits, pieces of data," he said. "A denial-of-service attack is the exact opposite of that. It is taking one person's will and imposing it on a bunch of others." In any case, Lyon added, his clients now included mainstream businesses—a Japanese game company, foreign-exchange traders, and a multibillion-dollar corporation that wanted to have additional security in the days before its IPO

Lyon first gained a measure of online fame in 2003, with a project called Opte, in which he created a visual map of the entire Internet—its backbone, transfer points, major servers. After reading that a similar project had taken several months to complete, he bet a friend that he could do it in a day, and he won. (A gorgeously rendered print of the map—which Lyon licenses free of charge—appeared in a traveling exhibition on the future of design.)

Lyon's obsessive interest in computer networks began early. In the third grade at a Sacramento, California, private school for learning-disabled children—Prolexic's name derives from Lyon's pride in overcoming severe dyslexia—he and a friend hacked a simple computer game. In junior high school, Lyon discovered the Internet, and with a friend, Peter Avalos, he soon founded a company called

TheShell.com, which provided accounts to chat-room users. But his grades suffered, and, after high school, he failed a year's worth of classes at California State University at Chico.

When a friend he met online, Robert Brown, offered Lyon a job at his computer-security company, Network Presence, he quit school and took it. Brown sent him off to secure the network of a large insurance company in the Midwest. Lyon was 19, and, he said, "I looked 13. So I wore a suit every day, and I worked my ass off for those guys." He burned out after two years—"I didn't know you had to meter yourself"—and returned to school, this time at California State University at Sacramento. There, Lyon signed up for philosophy classes, dumped his computers in a closet, and joined the rowing team. But he couldn't get away from computers entirely; he still took assignments from his old employer, and he and Avalos (who graduated from the United States Naval Academy and has recently returned from flying P-3s in Iraq) continued to operate TheShell.com. The company's clients tended to be advanced Internet users, and this had the effect of bringing the site to the attention of hackers. At one point, Lyon was fighting off several zombie attacks a day.

In August 2002, Dana Corbo, the CEO of Don Best Sports, called Network Presence for help. Don Best, which is based in Las Vegas, is a kind of Bloomberg for the gambling world, providing betting lines for both real-world and online casinos. The company had ignored an e-mailed extortion demand for $200,000, and it was under attack. Network Presence sent Lyon.

The next day, Lyon and another engineer flew to Las Vegas and helped Don Best's engineers set up powerful new servers. Lyon's strategy worked: the attackers gave up. Corbo treated them to a night out in Vegas, with dinner in front of the Bellagio fountains. (He also paid Network Presence a fee.)

Lyon still wanted to find out who was behind the attacks. He and Brown scanned the traffic data; found a zombie; and, thanks to an opening in Microsoft Windows, were able to see what other computers it had been connected to. This led them to a chat server in Kazakhstan; when they connected to it, they saw more attacks in progress. They notified the FBI and the Secret Service, but, Brown said, "they sort of threw up their arms, because it was in Kazakhstan." To Lyon, however, the lesson was clear: With clever techniques and a little luck, any attacker could be found.

In the late spring of 2003, Mickey Richardson, the general manager of Betcris, a Costa Rican–based gambling firm, received an extortion e-mail. (Online bookmaking, which is illegal in the United States, has flourished in Costa Rica and the Caribbean since the mid-1990s.) The letter requested $500 in eGold—an online currency—and was followed by an attack that crippled Betcris's Web site, its main source of revenue. Richardson couldn't afford to have the site disabled. He paid the $500.

The extortionists began hitting other offshore bookmakers. One firm after another paid up, anywhere from $3,000 to $35,000, which they wired to addresses in Russia and Latvia. Richardson expected that he, too, would be hit again. He heard about Don Best's successful defense and called Lyon. But Lyon was back in school and reluctant to take the job. Instead, he told Richardson to buy a server that was specially designed to filter out attacks. "The box," as Richardson called it, cost about $20,000. Over the phone, Lyon helped Richardson's information-technology manager, Glenn Lebumfacil, configure it.

A few months later, Richardson got another e-mail

from the extortionists. It arrived just before Thanksgiving, one of the busiest betting periods of the year, and it asked for $40,000. The e-mail said:

> If you choose not to pay for our help, then you will probably not be in business much longer, as you will be under attack each weekend for the next 20 weeks, or until you close your doors.

Richardson believed that he had "everything in place to protect the store," and he refused to pay.

When the attack came, it took less than 20 minutes to overwhelm the box. The data flood brought down both Betcris and its Internet service provider. After a few days of trying in vain to make the box work, Lebumfacil called Lyon in a panic. "Hey, man, remember that thing you set up for us?" he said. "It just got blown away."

Lyon saw a business opportunity. He quit school again and started a company, with Betcris as his first customer. He knew that he couldn't just add capacity to Betcris's system to capture the zombies, as he had with Don Best, because Costa Rica wasn't wired for that sort of system—there wasn't enough capacity in the entire country. So he decided to build his own network in the United States and use it to draw the attackers away from Betcris. The extortionists would think they were attacking a relatively defenseless system in Central America but would find themselves up against Lyon's machines instead.

Richardson, meanwhile, was stalling for time with the extortionists, claiming a medical emergency. "I guess you did not take my warning seriously," came the reply. "The excuse that you were in the hospital does not matter to me." The correspondence became increasingly belligerent. "Sorry moron but I am just having so much fun fucking with you," one e-mail said, raising the price to $60,000. Richardson responded by offering the extortionists jobs in Betcris's IT department. "I appreciate the offer to do work for you, but we are completely booked until the football season is over," one of them replied.

As Lyon brought his system online, the confrontation turned into a chess match. "Every time Barrett would change something, these guys would change something else," Brian Green, the CEO of Digital Solutions, Betcris's Internet service provider, said. "They threw wrenches, they threw everything they could at Betcris."

Finally, after three weeks, the attacker gave up. "I bet you feel real stupid that you did not keep your word," he wrote. "I figure by now you have lost 5 times what we asked and by the end of the year your decision will cost you more than 20 times what we asked." Richardson says that those numbers may not have been far off.

By then, everyone in the insular gaming world seemed to have heard that Lyon could stop zombie attacks, and he was getting calls from Jamaica, Costa Rica, and Panama. "It was kind of like stumbling into this strange little community in the middle of nowhere, where everybody worships a weird stone," Lyon said. "They all had superstitions about when they were going to be attacked."

Lyon decided, once again, to trace the source of the attack. He and Dayton Turner, a goateed 24-year-old engineer he had hired, allowed one of their own machines to become a zombie and watched as it was drawn into the botnet; by early January they had found the chat channel that controlled the zombies.

Logging on as "hardcore," Turner pretended to be a bot herder who had been out of the game for a while. "i want to get back into it," he wrote. "i ha[v]e a small group of zombies so far which is why i came back looking." Turner had spent years in chat rooms and communicated easily in the emoticon-heavy shorthand common to hackers. He gradually ingratiated himself with a Russian who called himself eXe and often logged in from a server that he'd named "exe.is.wanted.by.the.FBI.gov." Other members were not so welcoming; when Turner wrote, "i wanna help," one of them, uhdfed, replied, "we don't need ur HELP," and set his zombies on him. But Lyon and Turner kept returning, establishing their technical credibility and becoming a part of the scene.

They continued the ruse for weeks, occasionally with an FBI agent on the phone helping to direct the conversation. As bait, Turner described a program he had written that would help eXe to collect zombies, which he promised to give him as soon as he could rewrite it in a different programming language. "It was a matter of simply befriending the guy and making him think that he could trust us," Lyon said. Piece by piece, eXe revealed himself:

> hardcore: its pretty cold here right now, what's russia like? hehe
> eXe: i'm good
> eXe: something hot
> eXe: =)
> eXe: Russia is like the Russian Vodka=)
> hardcore: hehehe
> eXe: u give me code?

At one point, during an exchange about the number of computers each had infected, eXe asked Turner how old he was. Turner replied that he was 23 and added, "How about you? :)." eXe told him that he was a 21-year-old Russian student named Ivan. Turner said that his name was Matt and he lived in Canada. Then, trying to provoke a confession, he told Ivan that he made money from extortion: "They always pay because they want their business back and they don't want to admit they have a weakness . . . stupid Americans."

Turner then asked Ivan about a specific attack: "I figured it would be you since you have so many bots :P."

"Good idea . . . hehe," Ivan replied.

Before they signed off, Ivan wrote, "Bye friend."

In February 2004, Lyon and Turner submitted a 36-page report to the FBI and the NHTCU, outlining their profile of Ivan and their correspondence with his crew. At this point, they were operating as DigiDefense International, which Lyon had founded, hiring Turner and Lebumfacil as his first employees. At the company's temporary headquarters, in an office building in Costa Rica, paranoia about reprisals from Russian mobsters reigned, even though there were armed guards in the lobby. Meanwhile, Lyon and Turner kept chatting with Ivan.

A few weeks later, on a Saturday in March, Ivan slipped up: he logged in to the chat room without disguising his home Internet address. The same day, Turner happened to be online and decided to look up eXe's registration information. To his astonishment, he found what appeared to be a real name, address, and phone number: Ivan Maksakov, of Saratov, Russia. Lyon dashed off an e-mail to the authorities with the subject line "eXe made a HUGE mistake!"

A few months later, the Russian police, accompanied by agents from the NHTCU, swept into Maksakov's home, where they found him sitting at his computer. In television footage of the arrest, Maksakov looks like a clean-cut kid, with brown hair and a teenager's face. He sits glumly on his bed in shorts and a T-shirt as the police rummage through his room and carry out his equipment. The video shows the officers walking him to the local station and slamming the door shut on his cell.

In simultaneous raids in St. Petersburg and Stavropol, the police picked up four other Russians whom the NHTCU had traced by setting up a sting at a bank in Riga, Latvia, where a British company that was cooperating with the authorities had been directed to send its payment. "We were waiting for people to come pick the money up," Mick Deats, of the NHTCU, told me. "But that didn't happen immediately. What did happen was that the bad guys we were watching picked up lots of different payments—not ours. We were seeing them pick up Australian dollars, U.S. dollars, and denominations from all over the world. And we're thinking, Whose money is that?"

The NHTCU has never explicitly credited Prolexic's engineers with Maksakov's arrest. "The identification of the offenders in this came about through a number of lines of inquiry," Deats said. "Prolexic's was one of them but not the only one." In retrospect, Lyon said, "The NHTCU and the FBI were kind of using us. The agents aren't allowed to do an Nmap, a port scan"—techniques that he and Dayton Turner had used to find Ivan's zombies. "It's not illegal; it's just a little intrusive. And then we had to yank the zombie software off a computer, and the FBI turned a blind eye to that. They kind of said, 'We can't tell you to do that—we can't even suggest it. But if that data were to come to us we wouldn't complain.' We could do things outside of their jurisdiction." He added that although his company still maintained relationships with law-enforcement agencies, they had grown more cautious about accepting help.

When the authorities picked up Ivan Maksakov, he was one semester away from graduation at a technical college in Saratov. He spent five months in prison before being released on bail and now awaits trial. According to the authorities, he was a lower-level operative in the gang, which paid him about $2,000 a month for his services. A source close to the investigation told me that Maksakov, who faces 15 years in jail, is cooperating with the Russian police.

One afternoon in Prolexic's offices, I asked Turner if he had felt a sense of justice when Ivan was arrested. "I suppose," he said halfheartedly. "It was a difficult situation for me when I saw his picture, because I kind of felt for the kid. He wasn't necessarily a bad kid." Perhaps, Turner told me, Ivan had "just said, 'Let's see if it works. Hey, it works, and people pay me for it.'"

Lyon, too, was one semester from graduation when he dropped out of college to start his company. He was, in his own way, unable to resist the challenge, and he, too, had discovered that people would pay him for what he did. I asked him if he'd ever done anything illegal on the Net. He thought for a minute and then told me that once, as a teenager, he had poked around and discovered a vulnerability at Network Solutions, the company that at the time registered all the Web's addresses. "I went in and manipulated some domain names," he said. "A month later, I got a call from somebody with a badge," who had traced the intrusion back to Lyon's computer.

In the end, Lyon said, the authorities let it go. Those were simpler times. "I was scared shitless, but I learned my lesson," he said. "If something like that happened now, I can't imagine what would happen to me."

About the Contributors

David A. Bell is Andrew W. Mellon Professor in the Humanities at Johns Hopkins University and a contributing editor of the *New Republic.* He has held fellowships from the Guggenheim Foundation, the American Council of Learned Societies, the National Endowment for the Humanities, and the Woodrow Wilson International Center for Scholars. He writes widely on European history and politics. His book *The First Total War* will be published by Houghton Mifflin in early 2007.

David Bernstein is a senior editor at *Chicago* magazine. Previously he was a freelance writer, frequently contributing to the *New York Times, Chicago,* and *Crain's Chicago Business.* David has also been featured on WBEZ Chicago Public Radio. He lives in Chicago.

Mike Daisey is a monologuist, author, and professional dilettante. He is the author of the memoir *21 Dog Years: Doing Time @ Amazon.com,* and his new monologue, *Great Men of Genius,* explores the lives of Bertolt Brecht, P. T. Barnum, Nikola Tesla, and L. Ron Hubbard as a way of viewing the idea of genius. He can be found on the Web at mikedaisey.com and lives in Brooklyn with his wife and director, Jean-Michele Gregory.

Joshua Davis is a contributing editor at *Wired.* His story "La Vida Robot" generated so much reader interest in the Carl Hayden Robotics Team that a scholarship was formed and over $80,000 was raised to send Oscar, Cristian, Lorenzo, and Luis to college. Hollywood also took notice—Warner Brothers optioned the rights to the story, and John Wells, the creator of *ER* and *The West Wing,* has signed on to produce. Josh is an executive producer on the project. More information on the La Vida Robot scholarship can be found at www.joshuadavis.net.

Jay Dixit is senior editor at *Psychology Today.* He's written for the *New York Times,* the *Washington Post, Wired, Slate, Salon, Legal Affairs,* and *Rolling Stone,* covering technology, behavioral science, and cultural trends. A native of Ottawa, Canada, he studied at Yale University. He lives in Brooklyn.

Daniel Engber writes the "Explainer" column for *Slate.* He is a frequent contributor to National Public Radio, and his written work has appeared in *Salon, Seed, Popular Mechanics,* and the *Chronicle of Higher Education.* In 2005, his idea for how to distract basketball free-throw shooters was selected for the *New York Times Magazine*'s "Year in Ideas" issue.

Dan Ferber is a contributing correspondent for *Science,* and his articles about science, technology, health, and the environment have appeared in *Reader's Digest, Popular Science, Audubon,* and *Sierra.* He enjoys writing about scientists and their unusual obsessions. He lives in Indianapolis.

Steven Johnson is the best-selling author of *Everything Bad Is Good for You: How Today's Popular Culture Is Actually Making Us Smarter; Mind Wide Open: Your Brain and the Neuroscience of Everyday Life; Emergence: The Connected Lives of Ants, Brains, Cities, and Software;* and *Interface Culture: How New Technology Transforms the Way We Create and Communicate.* He writes the "Emerging Technology" column for *Discover* magazine, is a contributing editor to *Wired,* writes for *Slate* and the *New York Times Magazine,* and lectures widely. His fifth book, *The Ghost Map,* will be published by Riverhead in October 2006. Johnson lives in New York City with his wife and their two sons.

Steven Levy is a senior editor at *Newsweek,* where he writes a column called "The Technologist." His articles, opinion pieces, and reviews have appeared in a wide range of publications, including the *New Yorker,* the *New York Times Magazine, Harper's, Rolling Stone, Premiere,* and *Wired,* and have received numerous awards. Levy has also written five books, including *Hackers,* which *PC Magazine* named the best sci-tech book written in the last 20 years, and *Crytpo,* which won the grand eBook prize at the 2001 Frankfurt Book Festival. His sixth book, *The Perfect Thing,* about Apple's iPod music player, will be published by Simon & Schuster in October 2006. He lives in New York City and western Massachusetts with his wife, Pulitzer Prize–winning journalist and author Teresa Carpenter, and their son.

Farhad Manjoo is a reporter at *Salon* who first became interested in the Google Book Search story for its fascinating collision between copyright law—which had been erected to protect authors—and a new technology that will so clearly benefit everyone in society, including authors. Here is a clear instance where copyright law is hindering cultural advance. He hopes to show people exactly why they might need to change the law.

Lisa Margonelli is an Irvine Fellow at the New America Foundation. Her book, *Oil on the Brain: Travels in the World of Petroleum,* will be published by Nan A. Talese/Doubleday in December 2006. She works in Oakland, California.

David McNeill is a Tokyo-based journalist and university teacher. He is the Japan correspondent for the *London Independent* newspaper and a regular contributor to the *Irish Times, Japan Times,* and other publications and radio stations. He runs a mile from karaoke machines.

Justin Mullins is an artist and science writer based in London. His artwork is at www.justinmullins.com, and he is a consultant editor at *New Scientist* magazine.

Koranteng Ofosu-Amaah is a software engineer working on collaboration software and Web technologies at Lotus/IBM. From Ghana by way of France and England, he is an electrical engineer by training courtesy of Harvard University. He writes extensively on Africa, music, literature, technology, politics, and more. In his writing, he aims to celebrate the small things and demystify them in the spirit of a cultural interpreter. His current focus is on finding ways to move the software world through its industrial revolution by adopting the Web style.

Adam L. Penenberg is a journalism professor at New York University and assistant director of the Business & Economic Reporting Program. In 1998, while a staff editor at *Forbes.com,* he garnered national attention for unmasking Stephen Glass as a fabulist, as portrayed in the 2003 film *Shattered Glass* (Steve Zahn plays Penenberg). His first book, *Spooked: Espionage in Corporate America* (Perseus Books, 2000), was excerpted in the *New York Times Sunday Magazine,* and his second, *Tragic Indifference: One Man's Battle with the Auto Industry over the Dangers of SUVs* (HarperBusiness, 2003), was optioned for the movies by Michael Douglas. A former columnist for *Slate* and *Wired News,* Adam is currently a contributing writer for *Fast Company* magazine.

Daniel H. Pink is the best-selling author of *A Whole New Mind* and *Free Agent Nation.* He is a contributing editor at *Wired* and a columnist for Yahoo! Finance. His articles on business, technology, and economic transformation have also appeared in the *New York Times, Harvard Business Review, Fast Company,* and other publications.

Evan Ratliff is a freelance writer and the coauthor of *Safe: The Race to Protect Ourselves in a Newly Dangerous World* (HarperCollins, 2005). A contributing editor at *Wired* magazine and a contributing writer at *ReadyMade* magazine, he has also written for the *New Yorker, Outside,* the *New York Times,* and other publications. He lives in San Francisco.

Alex Ross has been the music critic of the *New Yorker* since 1996. His first book, *The Rest Is Noise: Listening to the Twentieth Century,* will be published in the fall of 2007 by Farrar, Straus, and Giroux.

Jim Rossignol is obsessed with games. In the last decade he has been channeling this obsession into writing for magazines and Web sites such as *Wired, PC Gamer, Gamasutra,* and the *London Times.* He lives in a crumbling stone house in the southwest of England, where he carefully researches the future. Elements of this research can be found at www.big-robot.com.

Jesse Sunenblick is a Brooklyn-based writer of journalism and fiction and a contributing writer to the *Columbia Journalism Review.* He is currently writing a book about baseball in Latin America. He habitually listens to Red Sox games, by radio, over the Internet.

Edward Tenner is a best-selling writer and speaker whose books include *Why Things Bite Back: Technology and the Revenge of Unintended Consequences* and *Our Own Devices: The Past and Future of Body Technology.* His essays and reviews have appeared in many leading magazines and newspapers in the United States and the United Kingdom, and his books have been translated into Japanese, Chinese, German, Italian, Portuguese, and Czech. Tenner is a member of the editorial board of *Raritan Quarterly Review* and a contributing editor of the *Wilson Quarterly* and *Harvard Magazine.* He is a senior research associate at the Lemelson Center for the Study of Invention and Innovation at the National Museum of American History, a visiting fellow at the Department of the History and Sociology of Science at the University of Pennsylvania, and an affiliate of the Center for Arts and Cultural Policy Studies. He lives in Plainsboro, New Jersey.

Clive Thompson writes about technology and culture for the *New York Times Magazine, Wired, New York Magazine,* and other publications. He also runs the tech-culture blog collision detection.net.

Joseph Turow is a professor at the University of Pennsylvania's Annenberg School for Communication. His newest book, published by MIT Press, is *Niche Envy: Marketing Discrimination in the Digital Age*.

Richard Waters is the San Francisco based West Coast editor and most senior technology correspondent for the *Financial Times.* He previously worked for the newspaper and its online edition, *FT.com,* on the "other" coast, where he was New York bureau chief and the paper's first information industries editor, as well as on the other side of the Atlantic, where he specialized in finance and capital markets. He is an unashamed BlackBerry addict.

Acknowledgments

Grateful acknowledgment is made to the following authors, publishers, and journals for permission to reprint previously published matierals.

"The Bookless Future: What the Internet Is Doing to Scholarship" by David A. Bell. First published in the *New Republic,* May 2 and 9, 2005. Reprinted by permission of the *New Republic,* © 2005, The New Republic, LLC.

"When the Sous-Chef Is an Inkjet" by David Bernstein. First published in the *New York Times,* Ferbruary 3, 2005. Copyright © 2005 by The New York Times Co. Reprinted with permission.

"The Coil and I: Adventures with a Mad Scientist's Lightning Machine" by Mike Daisey. First published in *Slate,* November 2, 2005. Copyright © by Mike Daisey. Reprinted by permission of the author.

"La Vida Robot" by Joshua Davis. First published in *Wired,* April 2005. Copyright © by Joshua Davis. Reprinted by permission of the author.

"Cats with 10 Lives" by Jay Dixit. First published in *Legal Affairs,* January/February 2005. Copyright © by Jay Dixit. Reprinted by permission of the author.

"Crying, While Eating: My Sad, Hungry Climb to Internet Stardom" by Daniel Engber. First published in *Slate,* June 23, 2005. Copyright © by Daniel Engber. Reprinted by permission of the author.

"Will Artificial Muscle Make You Stronger?" by Dan Ferber. First published in *Popular Science,* July 2005. Copyright © by Dan Ferber. Reprinted by permission of the author.

"Why the Web Is Like a Rain Forest" by Steven Johnson. First published in *Discover,* October 2005. Copyright © by Steven Johnson. Reprinted by permission of the author.

"The Trend Spotter" by Steven Levy. First published in *Wired,* October 2005. Copyright © by Steven Levy. Reprinted by permission of the author.

"Throwing Google at the Book" by Farhad Manjoo. First published in *Salon,* 2005. Copyright © by Salon. Reprinted by permission.

"China's Next Cultural Revolution" by Lisa Margonelli. First published in *Wired,* April 2005. Copyright © by Lisa Margonelli. Reprinted by permission of the author.

"Mr. Song and Dance Man" by David McNeill. First published in *Japan Focus,* September 22, 2005. Copyright © by David McNeill. Reprinted by permission of the author.

"Ups and Downs of Jetpacks" by Justin Mullins. First published in *New Scientist,* October 1, 2005. Copyright © by *New Scientist.* Reprinted by permission of *New Scientist.*

"Cultural Sensitivity in Technology" by Koranteng Ofosu-Amaah. First published in Koranteng's Toli, April 3, 2005. Copyright © by Koranteng Ofosu-Amaah. Reprinted by permission of the author.

"The Right Price for Digital Music: Why 99 Cents per Song Is Too Much, and Too Little" by Adam L. Penenberg. First published in *Slate,* December 5, 2005. Copyright © by Adam L. Penenberg. Reprinted by permission of the author.

"The Book Stops Here" by Daniel H. Pink. First published in *Wired,* March 2005. Copyright © by Daniel H. Pink. Reprinted by permission of the author.

"The Zombie Hunters: On the Trail of Cyberextortionists" by Evan Ratliff. First published in the *New Yorker,* August 10, 2005. Copyright © by Evan Ratliff. Reprinted by permission of the author.

"The Record Effect" by Alex Ross. First published in the *New Yorker,* June 6, 2005. Copyright © by Alex Ross. Reprinted by permission of the author.

"Sex, Fame, and PC Baangs: How the Orient Plays Host to PC Gaming's Strangest Culture" by Jim Rossignol. First published in *PC Gamer UK,* April 2005. Copyright © by Jim Rossignol. Reprinted by permission of the author.

"Into the Great Wide Open" by Jesse Sunenblick. First published in *Columbia Journalism Review,* January/February, 2005. Copyright © by Jesse Sunenblick. Reprinted with permission of the author.

"The Rise of the Plagiosphere" by Edward Tenner. First published in *Technology Review* in June 2005. Copyright © by Edward Tenner. Reprinted by permission of the author.

"The Xbox Auteurs" by Clive Thompson. First published in the *New York Times Magazine,* August 7, 2005. Copyright © by Clive Thompson. Reprinted by permission of the author.

"Have They Got a Deal For You: It's Suspiciously Cozy in the Cybermarket" by Joseph Turow. First published in the *Washington Post,* June 19, 2005. Copyright © by Joseph Turow. Reprinted by permission of the author.

"Plugged Into It All" by Richard Waters. First published in *Financial Times,* November 11, 2005. Copyright © by Richard Waters. Reprinted by permission of the author.

Every effort has been made to trace the ownership of all copyrighted material in this book and to obtain permission for its use.

Thanks are also due to Maria Bonn, Eszter Hargittai, Wendy Seltzer, Scott Shapiro, and Nick Thompson, all of whom provided invaluable advice and help throughout this project.

Text design by Mary H. Sexton

Typesetting by Delmastype, Ann Arbor, Michigan

The text font is Granjon, which was designed in 1928 for Linotype by George Jones using Claude Garamond's (1499–1561) late Texte (16 point) roman as his model. It is named after the sixteenth-century French printer, publisher, and lettercutter Robert Granjon (1513–89).

—*Courtesy adobe.com and myfonts.com*